TECHNICAL
REPORT

An Assessment of
the Ability of the U.S.
Department of Defense
and the Services to
Measure and Track
Language and Culture
Training and Capabilities
Among General
Purpose Forces

Jennifer DeCamp, Sarah O. Meadows, Barry Costa,
Kayla M. Williams, John Bornmann, Mark Overton

Prepared for the Office of the Secretary of Defense

Approved for public release; distribution unlimited

 NATIONAL DEFENSE RESEARCH INSTITUTE

The research described in this report was prepared for the Office of the Secretary of Defense (OSD). The research was conducted jointly by the MITRE Corporation's Department of Social, Behavioral, and Linguistic Sciences and the RAND National Defense Research Institute, a federally funded research and development center sponsored by OSD, the Joint Staff, the Unified Combatant Commands, the Navy, the Marine Corps, the defense agencies, and the defense Intelligence Community under Contract W74V8H-06-C-0002.

Library of Congress Cataloging-in-Publication Data

An assessment of the ability of U.S. Department of Defense and the services to measure and track language and culture training and capabilities among general purpose forces / Jennifer DeCamp ... [et al.].
   p. cm.
 Includes bibliographical references.
 ISBN 978-0-8330-7667-0 (pbk. : alk. paper)
 1. Soldiers—Education, Non-military—United States. 2. United States—Armed Forces—Officials and employees—Education. 3. Language and languages—Study and teaching—United States. 4. Military education—United States—Evaluation. 5. Cultural competence—Government policy—United States—Evaluation. 6. United States—Armed Forces—Personnel management—Evaluation. 7. United States. Dept. of Defense—Personnel management—Evaluation. I. DeCamp, Jennifer.
 U716.A88 2012
 407.1'5—dc23

                                                                              2012034524

The MITRE Corporation is a not-for-profit organization that provides systems engineering, research and development, and information technology support to the government. It operates federally funded research and development centers for the Department of Defense, the Federal Aviation Administration, the Internal Revenue Service and Department of Homeland Security, and the Administrative Office of the U.S. Courts, with principal locations in Bedford, Mass., and McLean, Va. To learn more, visit: www.mitre.org.

The RAND Corporation is a nonprofit institution that helps improve policy and decisionmaking through research and analysis. RAND's publications do not necessarily reflect the opinions of its research clients and sponsors.

Published 2012 by the RAND Corporation
1776 Main Street, P.O. Box 2138, Santa Monica, CA 90407-2138
1200 South Hayes Street, Arlington, VA 22202-5050
4570 Fifth Avenue, Suite 600, Pittsburgh, PA 15213-2665
RAND URL: http://www.rand.org/
To order RAND documents or to obtain additional information, contact
Distribution Services: Telephone: (310) 451-7002;
Fax: (310) 451-6915; Email: order@rand.org

Preface

The purpose of this research was to assess the ability of the U.S. Department of Defense (DoD) to measure and track language, regional expertise, and culture (LREC) capabilities and training among general purpose forces (GPF). The research tasks addressed four specific questions:

1. According to the best available data, what is the relevance of LREC training and capabilities to overall unit readiness and mission accomplishment?
2. How does DoD currently track the LREC training and capabilities of GPF?
3. Does this tracking adequately reflect unit readiness and the ability to accomplish missions?
4. If not, how can DoD improve tracking of LREC training and capabilities to adequately reflect unit readiness?

To address these questions, the study team reviewed DoD policies and directives and the available academic literature, conducted interviews of practitioners and policymakers, and analyzed available survey data. This report presents the results of the study. This research should be of interest to policymakers interested in LREC skills and training, as well as those interested in readiness requirements.

This research was sponsored by the Defense Language Office (DLO) within the Office of the Under Secretary of Defense for Personnel and Readiness (OUSD[P&R]) and conducted jointly by the MITRE Corporation Department of Social, Behavioral, and Linguistic Sciences and within the RAND Corporation Forces and Resources Policy Center of the RAND National Security Research Institute, which are federally funded research and development centers (FFRDCs).

For more information on MITRE Corporation, see http://www.mitre.org.

The RAND National Defense Research Institute is an FFRDC sponsored by the Office of the Secretary of Defense, the Joint Staff, the Unified Combatant Commands, the Navy, the Marine Corps, the defense agencies, and the defense Intelligence Community.

For more information on the RAND Forces and Resources Policy Center, see http://www.rand.org/nsrd/ndri/centers/frp.html or contact the director (contact information is provided on the web page).

Contents

Figures

Tables

Summary

The Defense Language Office (DLO) tasked MITRE Corporation and the RAND National Defense Research Institute (NDRI) at the RAND Corporation, two federally funded research and development centers (FFRDCs), to jointly address questions concerning the U.S. Department of Defense's (DoD's) ability to measure and track the language, regional expertise, and culture (LREC) training and capabilities of general purpose forces (GPF). Using interviews with LREC practitioners and policymakers, a policy review, an academic literature review, and an analysis of survey data, the report addresses the following four questions. A more extensive summary is provided in Chapter Five of this report.

Research Questions

1. According to the Best Available Data, What Is the Relevance of Language, Regional Expertise, and Culture Training and Capabilities to Overall Unit Readiness and Mission Accomplishment?

Most LREC practitioners and policymakers interviewed for this study expressed the opinion that LREC capabilities are critical to the readiness and effectiveness of some units performing specific missions. Many interviewees argued that LREC capabilities are needed but that each unit or individual does not necessarily need the same types and mixes of skills as every other unit or individual. Because many of the required LREC skills differ by mission and task, it is critical to measure LREC mission readiness at the mission and task levels.

A substantial amount of anecdotal evidence, as well as policy, implies that LREC skills are essential to mission effectiveness. However, there have yet to be any rigorous, formal studies of whether receipt of LREC training is linked to improved individual job performance or unit mission accomplishment in the military. Several efforts have begun the process of collecting data, which could be used to assess such a relationship, through surveys (e.g., the Status of Forces Survey of Active-Duty Members [SOF-A] and a recent endeavor by the U.S. Marine Corps to interview redeploying marines).

A small but growing research literature uses data from service members with field experience to assess their perceptions of whether LREC training and skills are associated with job performance. Although these studies suggest that some basic knowledge (e.g., common words and phrases, local norms and customs, appreciation of foreign cultures) is useful, they do not indicate that these skills are essential for successfully performing one's job. Further, this line of research does not establish a causal link between LREC and mission effectiveness.

2. How Does the U.S. Department of Defense Currently Track Language, Regional Expertise, and Culture Training and Capabilities of General Purpose Forces?

The Language Readiness Index (LRI) provides systematic tracking of DLPT scores and self-reporting on language capabilities. More information will be added in 2012 regarding regional and cultural expertise. However, people interviewed for this study expressed concern that determining language, regional expertise, and cultural readiness—particularly for GPF, who may have very low levels of proficiency—may require additional sources of information. Not everyone with foreign-language skills receives FLPB or takes the DLPT or even the tests for very low proficiency. Not everyone who completes a self-reporting questionnaire fills out all information. Cultural and regional training is often done by the unit and is often tracked only at the unit level (i.e., the commander assumes that all members of his or her unit have completed the training, but there is rarely tracking of the information by individual so that, when the soldier joins a different unit, the new commander will know whether the soldier has had the training).

Aside from DLPT data, there is little tracking of other LREC capabilities among GPF. Some potentially relevant data (e.g., use of the language at home as a heritage speaker) are collected (e.g., at entry into a service, in Defense Language Institute Foreign Language Center [DLIFLC] interviews) but are not reliably tracked or available to unit commanders. As many interviewees pointed out, such data as home use of languages, university study, and prior deployments may indicate language and cultural or regional expertise, but currently no existing data allow for any reliable predictions of expertise.

Some types of language-related data are being pulled into the Language Readiness Index (LRI). The LRI is an effort currently being undertaken by DLO to create a combined database for tracking language qualifications of individuals across all services. However, in its current form, it does not include data on regional expertise or culture. It is not clear when the LRI will be fully operational, and it will likely undergo revisions as users provide real-time feedback.

3. To What Extent Does This Tracking Reflect Unit Readiness and the Ability to Accomplish Missions?

All interviewees indicated that tracking of language and culture is insufficient to adequately reflect unit readiness and the ability to accomplish missions. For DLO's purposes, it would be most useful to have a DoD-wide recommendation with respect to what constitutes mission readiness for GPF, similar to requirements for weapon training, equipment, and other readiness factors. Unfortunately, despite numerous concurrent studies about various aspects of LREC training and capabilities, making a "readiness recommendation" at this juncture would be premature and speculative at best. More analysis is needed.

4. How Can the U.S. Department of Defense Improve Tracking of Language, Regional Expertise, and Culture Training and Capabilities to Adequately Reflect Unit Readiness?

The recommendations in this report are based on policy, directives, and related academic literature, as well as the opinions provided in the interviews with LREC practitioners and policymakers. Both short- and long-term recommendations for improving LREC tracking and assessing the linkage between LREC training and skills and mission readiness and success are presented. The ultimate end goal of the recommended activities is to develop a set of readiness metrics, both at the general level for all GPF and at the mission-specific level. If LREC training is designed to bring the GPF to a very low but operationally effective level of training, there must be a mechanism in place to track that training, those skills, and the resulting operational effectiveness of the unit. None of those mechanisms is currently in place.

Recommendations

Short-Term Recommendations

Short-term recommendations include the following.

Recommendation 1: Standardize Language, Regional Expertise, and Culture–Related Terms

There is currently confusion about the term *LREC*, which is sometimes used to mean all of language, regional expertise, and culture and sometimes used to mean courses satisfying a military directive. Developing a common vocabulary could be accomplished through discussions and agreements at management meetings and through providing the terminology via a DLO website that can be accessed by key stakeholders in the LREC world. These stakeholders include not only DoD and service-level representatives of the Defense Language Action Panel (DLAP) but also researchers who are conducting studies related to LREC.

Recommendation 2: Develop Measures of Mission Effectiveness

It is not clear what it means to be mission effective or to achieve mission success. Given that defining these terms is necessary before the linkage between LREC training and skills can be assessed, these terms should be operationalized in a way on which key stakeholders agree. Workshops, such as those being conducted by the Joint Chiefs of Staff (JCS), could help to identify assessment approaches and measures. One approach may be to utilize focus groups of experts from the military and perhaps also from foreign partners to determine success metrics.

There will likely not be a single measure that adequately reflects mission effectiveness, nor will the association between LREC training and skills and mission effectiveness be one-to-one. That is, many other factors will influence the course of missions and how successful they are, and any means to collect data should attempt to capture all these different pieces of information.

Recommendation 3: Develop Standardized Language, Regional Expertise, and Culture After-Action Reports to Assess the Link Between Language, Regional Expertise, and Culture Training and Capabilities and Mission Success and Effectiveness

Standardized LREC after-action reports (AARs) would collect data that could then be used to assess the association between LREC training and skills and mission success (as defined in recommendation 2). This information would be collected across services, locations, and mission types. Such a systematic data effort would provide quantitative data that could be analyzed to estimate the connection between LREC skills and capabilities and mission success. The standardized AAR should include fields pertaining to the mission (e.g., type, date, region or location), characteristics of the unit and its members (e.g., LREC training received, self-reported skills, heritage speakers), and metrics of mission effectiveness and success (e.g., commander ratings, objective measures).

Recommendation 4: Develop Standardized Surveys for Deployed and Recently Deployed General Purpose Forces

As we noted in recommendation 3, which is aimed at commanders, additional data that could be used to link LREC training and skills to mission readiness and success could come from recently deployed GPF service members.

Long-Term Recommendations

Long-term recommendations include the following.

Recommendation 5: Develop an Infrastructure for Language, Regional Expertise, and Culture Data

This infrastructure would facilitate obtaining information on, evaluating, and researching LREC expertise, including the impact of this expertise on unit readiness and mission effectiveness. The LRI is a significant step toward improving data accuracy and making collected data more widely available. Means of developing this infrastructure would include standardizing terms, improving data accuracy, improving guidelines for self-reporting, and providing guidelines and information to unit commanders, officers, and researchers on using these data. It also includes development of a website or other knowledge-management structure to make data, evaluations of data, and other related LREC studies available to key stakeholders, including researchers.

Recommendation 6: Develop a Causal Model Linking Language, Regional Expertise, and Culture to Mission Success

One way to develop a causal model is to develop a bottom-up process in which smaller units are related and linked to one another. Such an endeavor is designed to link specific types of training and skills to specific mission outcomes. While a more detailed data-collection program will initially be expensive and time-consuming to establish, it would better automate data collection in the future. It will also supply valuable data to training providers to enable more-tailored courses. In addition, such a program would help to enable better management and development of LREC skills throughout a service member's career in the military. This approach is similar to what is being done by the Marine Corps in tracking and training by task.

Extensive work has been conducted by the Irregular Warfare Capabilities-Based Assessment Campaign from 2007 to the present. It has issued a Joint Operating Concept (now being updated) on the doctrine, organization, training, materiel, leadership and education, personnel, and facilities (DOTMLPF) change recommendations (DCRs) or initial capability documents for irregular warfare, defeating terrorist networks, preparation of the environment, security force assistance, foreign internal defense, unconventional warfare, joint civil affairs, and theater and army military information support to operations. The majority of these documents list language, regional expertise, and cultural capabilities as capability gaps that affect mission accomplishment. The analysis goes down to mission-essential task and condition level. These capabilities are needed to perform the missions and form the basis for readiness assessments.

In a similar vein, the Joint Staff–led Capabilities-Based Requirement Identification Process (CBRIP), based on Universal Joint Tasks, requires the geographic combatant commands to identify the language and regional and cultural capabilities required for their missions. When approved by the Joint Requirements Oversight Council, these geographic combatant command–expressed needs go to the force providers as demand signals. The force providers determine necessary quantities to meet these demands and alter their manning documents accordingly. The goal is to be able to use the LRI to match force requirements to language capability inventory to track overall force readiness from a language perspective. Work continues on developing regional proficiency inventory tracking and to enable a readiness comparison.

The end goals of establishing a causal model are to (1) link LREC training and skills to mission success and (2) provide sufficient data to establish what it means to be "LREC-ready." Although the approach outlined in this report is bottom-up in that it builds on individual tasks, once established, it will lend itself to a higher-level recommendations about minimum levels of LREC capabilities. Extensive validation is needed.

Recommendation 7: Develop Tests of Training (i.e., Learning) That Are Associated with Skills That Have Been Linked to Mission Readiness

Recommendation 6 suggests that both tracking and training be linked to operational missions (e.g., task-based training) to link LREC to mission effectiveness. Also, as noted in recommendation 4, surveys of GPF either while in the field or after having just returned may provide valuable insight on the effectiveness of LREC training. Training and assessment should be simultaneously developed in order to ensure that students are taught knowledge, skills, and attitudes that are accurately reflected in testing and assessment mechanisms (i.e., students are not tested on things that were not taught).

Acknowledgments

We gratefully acknowledge all the individuals interviewed. Their time and thoughts were integral to this effort.

We also thank Beth J. Asch, J. Michael Polich, Chaitra M. Hardison, and Jennifer J. Li at RAND and Charlotte Evans at MITRE for their reviews and comments.

Abbreviations

3C	cross-cultural competency
AAR	after-action report
ADLS	Advanced Distributed Learning System
AETC	Air Education and Training Command
AFCLC	Air Force Culture and Language Center
AFRICOM	U.S. Africa Command
AKO	Army Knowledge Online
AOR	area of responsibility
ARI	U.S. Army Research Institute
ARNG	Army National Guard
ATRRS	Army Training Requirements and Resources System
CAOCL	Center for Advanced Operational Culture Learning
CCT	cross-cultural training
CENTCOM	U.S. Central Command
CI	confidence interval
COCOM	combatant command
COIN	counterinsurgency
COMISAF/USFOR-A	Commander, International Security Assistance Force/ Commander, U.S. Forces–Afghanistan
CRL	culture, region, and language
DA	Department of the Army
DLAP	Defense Language Action Panel
DLIFLC	Defense Language Institute Foreign Language Center
DLO	Defense Language Office

DLPT	Defense Language Proficiency Test
DMDC	Defense Manpower Data Center
DoD	U.S. Department of Defense
DoDD	U.S. Department of Defense directive
DRRS	Defense Readiness Reporting System
DTMS	Digital Training Management System
EST	Expeditionary Skills Training
EUCOM	U.S. European Command
FAO	foreign area officer
FFRDC	federally funded research and development center
FLO	Foreign Language Office
FLPB	Foreign Language Proficiency Bonus
FM	field manual
GAO	U.S. Government Accountability Office
GCC	geographic combatant commander
GPF	general purpose forces
HQ	headquarters
J1	Joint Chiefs of Staff Manpower and Personnel Directorate
J-3	operations directorate of a joint staff
J-5	plans directorate of a joint staff
JCS	Joint Chiefs of Staff
LREC	language, regional expertise, and culture
LRI	Language Readiness Index
MCIA	Marine Corps Intelligence Activity
METL	mission-essential task list
MOS	military occupational specialty
NCO	noncommissioned officer
NDRI	RAND National Defense Research Institute
OEF	Operation Enduring Freedom
OIF	Operation Iraqi Freedom

OPI	Oral Proficiency Interview
OSD	Office of the Secretary of Defense
OUSD(I)	Office of the Under Secretary of Defense for Intelligence
PACOM	U.S. Pacific Command
RCLF	Regional, Culture, and Language Familiarization
SCEMS	U.S. Southern Command Enterprise Management System
SecDef	Secretary of Defense
SOF-A	Status of Forces Survey of Active-Duty Members
SORTS	Status of Resources and Training System
SOUTHCOM	U.S. Southern Command
TDY	temporary duty
TRADOC	U.S. Army Training and Doctrine Command
UJTL	Universal Joint Task List
USD(P&R)	Under Secretary of Defense for Personnel and Readiness
USFOR-A	U.S. Forces–Afghanistan
USNORTHCOM	U.S. Northern Command
USSOCOM	U.S. Special Operations Command
VEST	Visual Expeditionary Skills Training
VLR	very low range

CHAPTER ONE

Introduction

The Defense Language Office (DLO) tasked MITRE Corporation and the RAND National Defense Research Institute (NDRI) at the RAND Corporation, two federally funded research and development centers (FFRDCs), to jointly address questions concerning the U.S. Department of Defense's (DoD's) ability to measure and track language, regional expertise, and culture (LREC) training and capabilities for general purpose forces (GPF).

Background

Government directives provide basic guidelines for GPF personnel with respect to LREC training (e.g., DoD Directive [DoDD] 1322.10 or counterinsurgency [COIN] training guidance). However, although these directives and guidelines are based in field expertise, there is little tracking from them to the specific mission requirements. Furthermore, there is concern that the current means of tracking such training and capabilities are incomplete or inconsistent and that they do not adequately reflect a unit's readiness or effectiveness in terms of mission success. Detailed specification, tracking, and validation are needed.

Objective and Research Questions

The objective of this task is to provide information to policymakers about the available data to track LREC training and skills, as well as available information on how LREC affects readiness and mission accomplishment.

To reach the stated objective, the following research questions were addressed:

1. According to the best available data, what is the relevance of LREC training and capabilities to overall unit readiness and mission accomplishment?
2. How does DoD currently track LREC training and capabilities of GPF?
3. To what extent does this tracking adequately reflect unit readiness and the ability to accomplish missions?
4. How can DoD improve tracking of LREC training and capabilities to adequately reflect unit readiness?

Scope and Organization of This Report

This project focuses on GPF. It does not include information relating to language professionals, such as translators, interpreters, and foreign area officers (FAOs), or relating to commands or agencies. It does include information relating to nonlanguage professionals deploying with National Guard and Reserve Components.

Chapter Two describes the methodology and data used in the study. Chapter Three addresses the first research question and uses available data to assess the importance of LREC training and skills for mission readiness and mission accomplishment. Chapter Four addresses the second research question and addresses how DoD currently tracks LREC training and skills and whether or not that tracking adequately reflects mission readiness. Finally, Chapter Five summarizes the findings and offers recommendations for linking LREC training and skills to mission readiness and success. In addition, we include four appendixes. Appendix A lists the policies and directives we reviewed for this analysis. Appendix B lists our interviewees, and Appendix C provides the interview questions we used. Appendix D details the confidence intervals (CIs) for our analysis of the Status of Forces Survey of Active-Duty Members (SOF-A).

CHAPTER TWO

Methodology and Data

To address the research questions outlined in Chapter One, different methodologies were used, including a review of relevant DoD and service LREC policies and directives and relevant research literature, a quantitative analysis of survey data, and qualitative interviews with LREC practitioners and policymakers. This chapter describes each analysis in more detail and notes how each is related to the research questions.

Review of Policies, Directives, and Academic Literature

The first step in addressing research question 1 (according to the best available data, what is the relevance of LREC training and capabilities to overall unit readiness and mission accomplishment?) involved reviewing LREC policies and directives from both DoD and the services (see Appendix A for a list of reviewed documents). This review allowed the research team to gauge the *subjective* (e.g., anecdotal) importance of LREC for mission readiness and accomplishment. The team searched these documents for examples that would indicate either a belief that LREC training or capability is important or actual quantitative assessment of the linkage between the LREC training or capabilities and mission success.

The research team did not define mission readiness or mission success or accomplishment a priori. This approach allowed the researchers to conduct a broad search for supporting evidence of the subjective and objective importance of LREC training and skills. However, one limitation of this approach is that the researchers did not have any evidence that indicates how well, if at all, these policies and directives are or were implemented.

The review also includes a discussion of the relevant literature from business and industry because it is common in multinational corporations for individuals to spend time abroad. This review allowed the team to gauge the *objective* (e.g., data analysis) importance of LREC for job and performance outcomes at both the individual and business levels. This literature can help the team to assess whether LREC training and skills are associated with business-relevant outcomes, which may or may not be relevant to the military.

Survey Data

To address the first research question, the team also utilized data from the June 2010 SOF-A (Defense Manpower Data Center [DMDC], 2010a, 2010b). A more detailed description of the data is provided in Chapter Three. The SOF-A, conducted for the DMDC, contains a subset of questions that ask service members about their experiences with predeployment language, region, and culture training. Specifically, the LREC data available from the SOF-A addressed seven characteristics of training: duration, trainer type, location, supplemental material, objectives, satisfaction, and job enhancement. These data also provide a view of the importance of LREC training from the perspective of deployed service members, as opposed to the interviews with LREC practitioners and policymakers (discussed in the next section).

The SOF-A is a web-based survey of active-duty service members across the Air Force, Army, Navy, and Marine Corps that provides the Under Secretary of Defense for Personnel and Readiness (USD[P&R]) with information on the attitudes and opinions of the force on many quality-of-life issues. Although comparable surveys are available for Reserve Component members and DoD civilian employees, the research team used data for the June SOF-A, conducted between June 11 and July 22, 2010, which focused only on active-duty members.

Of the 68,540 active-duty service members eligible to participate,[1] 18,983 returned usable surveys. The adjusted weighted response rate was 25.3 percent (see DMDC, 2010a). Unless otherwise stated, the results presented here are weighted to reflect overall population values (for a detailed explanation of the sampling frame and weights, see DMDC, 2010b, and DoD, 2002). Results are presented for all service members, as well as separately by officer versus enlisted status.

Interviews with LREC Practitioners and Policymakers

The selection of interviewees began with a list provided by DLO. Included in the list were three main groups of interviewees: (1) LREC *users*, who provide the demand signal for LREC skills and capabilities; (2) LREC *providers*, who respond to demand signals by delivering training to service members with training; and (3) organizations responsible for oversight and tracking of LREC matters. LREC users consisted of the geographic combatant commands (COCOMs) (e.g., U.S. Northern Command [USNORTHCOM], U.S. Central Command [CENTCOM]). LREC providers included the services (i.e., Air Force, Army, Marine Corps, and Navy) and the institutions directly responsible for training (e.g., the Defense Language Institute Foreign Language Center [DLIFLC]). Offices and organizations responsible for oversight and tracking of LREC included the DMDC and the Joint Chiefs of Staff (JCS).

The interview offices and organizations were contacted through their representatives on the Defense Language Action Panel (DLAP). At the end of each interview, the research team asked whether there were any other organizations or individuals whom they should interview about LREC tracking, training, skills, or capabilities of GPF. In total, the team interviewed LREC practitioners and policymakers from 18 different organizations—at many organizations, interviewing more than one individual. A full list of offices and organizations interviewed and the dates of the interviews are shown in Appendix B.

Interviews were semistructured and based on a predetermined set of questions aimed at addressing the four research questions outlined in Chapter One.[2] Separate sets of questions were designed for LREC providers (e.g., the services, DLIFLC) and LREC users (e.g., the

---

[1]   Only service members with at least six months of service and who are below flag rank were eligible to participate.

[2]   A copy of the interview questions is provided in Appendix C.

COCOMs). Oversight and tracking organizations received slightly modified versions of these two sets of questions.

One of the goals of the qualitative interviews was to determine how the services and COCOMs are currently using data to assess linkages between LREC requirements and accomplishment of mission tasks (e.g., research question 1). The following questions were used to address this issue:

- To what degree are LREC capabilities relevant to overall unit readiness?
- For what types of missions are LREC capabilities important?
- How can one determine whether LREC training and capabilities are effective in promoting mission readiness and accomplishment among GPF?

A second goal of the qualitative interviews was to determine how DoD currently tracks LREC training and capabilities and the degree to which this tracking accurately reflects readiness (i.e., research questions 2 and 3). The following questions were used to address this issue:

- How do you track LREC capabilities?
- How do you track LREC training?
- How well does current tracking of both training and capabilities reflect unit readiness?

Additional interview questions addressed LREC curriculum, who determines it and how, and documentation regarding LREC requirements, implementation of policy, measurement, and tracking.

CHAPTER THREE

Impact of LREC Training and Capabilities on Mission Readiness and Accomplishment

Although intuitively it appears that LREC skills should be causally linked to success in military operations—especially those that involve cultures different from one's own—the evidence supporting this assumption is sparse and, at best, anecdotal. There are many narratives about the usefulness of LREC skills, especially in certain types of military operations, and particularly among those that occurred as part of Operation Iraqi Freedom (OIF) and Operation Enduring Freedom (OEF). For example, in January 2011, General Sir David Richards commented that seeing *The Great Game,* a play based on Afghan history and culture, would have made him a better commander during his own deployment (Norton-Taylor, 2011). MG Michael Flynn, in *Fixing Intel: A Blueprint for Making Intelligence Relevant in Afghanistan* (Flynn, Pottinger, and Batchelor, 2010), discussed the importance of local knowledge, including cultural knowledge, for informing good decisions. He stated that this knowledge and intelligence is not collected by specialized collectors but by the GPF on the ground interacting with the local population. And LTC Dale Kuehl, a battalion commander in Iraq in 2007, oversaw operations and observed that the level of violence in his area of operations decreased during his deployment in response to effective COIN tactics. He concluded that "gaining the trust of the local populace was essential to our operations" (Kuehl, 2009, p. 72).

This belief that cultural awareness is important for mission accomplishment has been included in COIN doctrine: Army and Marine Corps Field Manual (FM) 3-24 states, "Successful conduct of COIN operations depends on thoroughly understanding the society and culture within which they are being conducted . . . effective COIN operations require a greater emphasis on certain skills, such as language and cultural understanding, than does conventional warfare" (Department of the Army, 2006, pp. 1-22, 1-23). Although expert opinion is a valuable source of data for assessments and analyses, there have yet to be any rigorous, formal studies of whether LREC capabilities are linked to improved individual job performance or unit mission accomplishment in the military. There is also little literature specifically addressing the question of LREC skills in a military context.

This rigorous formal study—as described in Chapter Five—is needed to scientifically demonstrate and measure that LREC has an impact on mission effectiveness. The ability to scientifically demonstrate and measure such impact is needed for responses to Congress. It is also needed to measure and compare the impact of practices in hiring and training to make decisions about the allocation of scarce resources. Given the competing demands of different types of training among GPF, it is important to have an assessment of, and justification for, the priority given to any given set of skills.

This chapter summarizes relevant literature and interview and survey data that support an association between LREC training and capabilities and unit readiness and mission accomplishment among GPF. It begins with a discussion of the guidance provided by DoD and the individual services regarding the importance of LREC skills. Some of the reviewed documents also contain information about training and therefore provide insight into the perceived importance of LREC.

Review of Policies and Directives

This section reviews the high-level policies and directives that have been issued regarding LREC in the past few years. Official guidance has increasingly stressed the importance of LREC. These documents reflect a belief that LREC is important. They also provide some indication of the goals of LREC training.

U.S. Department of Defense

Emphasis on the importance of LREC skills, as opposed to more-traditional, kinetic military skills, is common in experiences and lessons learned reported by service members coming out of OEF and OIF; however, they rarely provide specific examples, and there are no statistical data or formal studies. For example, the February 2005 *Defense Language Transformation Roadmap* stated,

> Post-9/11 military operations reinforce the reality that the Department of Defense needs a significantly improved organic capability in emerging languages and dialects, a greater competence and regional area skills in those languages and dialects, and a surge capability to rapidly expand its language capabilities on short notice. (DoD, 2005, p. 1)

Similarly, Chairman of the Joint Chiefs of Staff Instruction (CJCSI) 3126.01, *Language and Regional Expertise Planning* (Chairman of the Joint Chiefs of Staff, 2006) identified language skills and regional expertise as both "critical warfighting skills" and core competencies "integral to joint operations" (p. A-1). More-recent documents have also specified the types of operations in which LREC capabilities may be particularly relevant. DoD's 2010 *Quadrennial Defense Review Report* (DoD, 2010a) mentions LREC initiatives in two key mission areas: "succeed in counterinsurgency, stability, and counterterrorism operations" and "build the security capacity of partner states" (p. viii). It also cites building LREC expertise as an aspect of "developing future military leaders" (p. xiii).

DoD's 2011 *Strategic Plan for Language, Regional Expertise, and Cultural Capabilities 2011–2016* sets three primary goals that are reflective of the importance it places on LREC capabilities:

> Identify, validate, and prioritize requirements for language skills, regional expertise, and cultural capabilities, and generate accurate demand signals in support of DoD missions.

> Build, enhance, and sustain a Total Force with a mix of language skills, regional expertise, and cultural capabilities to meet existing and emerging needs in support of national security objectives.

> Strengthen language skills, regional expertise, and cultural capabilities to increase interoperability and to build partner capacity. (p. 8)

Services

The documents listed in the previous section give roughly equal weight to language skills, regional expertise, and cultural capabilities, and DoDD 5160.41E (2010b), Defense Language Program, established the following as policy: "Foreign language and regional expertise be considered critical competencies essential to the DoD mission." However, the services do not necessarily share those priorities: In general, the services emphasize cultural skills over language and regional expertise for GPF. Service missions differ, and mix of desired language skills, regional expertise, and cultural capabilities varies as well. At the time of this writing, the services are developing their approaches to regionally align elements of their forces. Regional alignment will affect current strategy documents as services reevaluate the language, regional, and cultural requirements for the GPF and a shift from cultural toward language emphasis during FY 2013 and beyond.

The services also use different terminology. For example, although the Navy uses the abbreviation *LREC*, the Air Force variously discusses cross-cultural competency (3C) and culture, region, and language (CRL) skills.

Army

The Army notes in its December 1, 2009, *Army Culture and Foreign Language Strategy* (Department of the Army [DA], 2009), "Battlefield lessons learned have demonstrated that language proficiency and understanding of foreign culture are vital enablers for full spectrum operations" (p. ii). Under the Army strategy, culture capability development is the main effort; language capability development is the supporting effort. In addition, the majority of the GPF are considered to require only a "rudimentary capability in culture and foreign language," but "a desired outcome . . . is for all leaders and Soldiers to achieve some level of proficiency in both culture and foreign language for some portion of the world" (p. iii).

The Rapport training program developed by DLIFLC is mandatory for all soldiers and DA civilians deploying to Iraq or Afghanistan, with content specific to each particular theater. The program includes six to eight hours of training, with a focus on ten military survival tasks. Students receive grades at the end of training, and, if a student successfully completes the course, his or her certificate of completion is automatically entered into the Army's Learning Management System.

DLIFLC also provides a HeadStart2 program for a "language and culture platoon level enabled leader." This is an 80- to 100-hour self-study course with an emphasis on language and should also be available through the Army Knowledge Online (AKO) website in the near future.

Air Force

In the May 2009 *Air Force Culture, Region and Language Flight Plan*, the Air Force refers to 3C as a "critical and necessary capability," while also calling CRL skills a "force enhancing capability" (U.S. Air Force, 2009, p. 2). The strategy is "3C for all Airmen and robust language skills and regional expertise for targeted Airmen" (p. 3). The CRL flight plan focuses on the importance of these skills for negotiating, communicating, and relating to both coalition partners and adversaries.

The Air Force Culture and Language Center (AFCLC) oversees culture training for deploying Air Force personnel. A primary source for culture training is through the online Advanced Distributed Learning Service (ADLS). This system, hosted by Air Education and

Training Command (AETC), provides online courses, including the culture general course, which is an introduction or general awareness course, and culture-specific courses, which address cultures of specific regions or countries. All of these courses are developed by the Expeditionary Skills Training (EST) division of AFCLC.

The culture general course is required of all airmen who are within a 12-month vulnerability period for deployment. It addresses understanding the 12 primary domains of culture so airmen can translate the information into usable tools to operate in a culturally diverse environment. This basic understanding is considered critical to mission success and survivability.

The culture-specific course is provided to airmen once they have received notification for deployment to a specific location, and the training is specific to that location. At the moment, there are culture-specific courses for Iraq and Afghanistan that are taught both online and in a classroom environment as part of the Air Advisor Course, as well as other expeditionary skill venues across the Air Force. Depending on the deployment location, more specifically tailored courses may be required before deployment. For example, an airman being deployed as an advisor to a local national unit (rather than in support of a U.S. unit) might also be required to take courses on cross-cultural communication, building cross-cultural relations, conflict management and negotiations, and language.

In addition to these courses, the AFCLC EST portal provides an Expeditionary Airman Field Guide to all deploying airmen. There are currently field guides for both Iraq and Afghanistan, and many of these have been delivered to service members outside the Air Force at Army and Marine Corps units at the request of those service members. EST has also produced an engaging training tool called the Visual Expeditionary Skills Training (VEST) project. The VEST project provides live-actor immersion training films that can be viewed by all service members from their desktop computers via the AFCLC website (AFCLC, 2012).

AFCLC is currently developing courses for locations beyond Afghanistan and Iraq. At the moment, AFCLC is developing courses for ten countries in U.S. Africa Command (AFRICOM) and two in U.S. Southern Command (SOUTHCOM). As other COCOMs express interest, the center will also develop courses for other countries.

Marine Corps

The *Marine Corps Language, Regional, and Culture Strategy: 2011–2015* (Headquarters, U.S. Marine Corps, 2011) discusses the importance of LREC training for both specialized units and GPF within the Marine Corps. For training GPF, the Marine Corps uses the *big C, little l* implementation plan, indicating the relative difference in emphasis on culture versus language skills. This plan primarily focuses on the operational cultural aspect of military action (e.g., an anthropologically driven model of how different cultural dimensions—environment, social, power, economy, belief systems, and political leadership—work together). Language instruction is targeted at specific functional tasks and aims to provide marines with tactical language skills.

The *Operational Culture and Language Training and Readiness Manual* (Department of the Navy, 2009) lays out LREC training tasks for inclusion in the overall Marine Corps training curriculum. The training tasks in the *Operational Culture and Language Training and Readiness Manual* are tied to the Core Capability Mission Essential Task List (METL), and can be tracked in both the Global Status of Resources and Training System (GSORTS) and the Defense Readiness Reporting System (DRRS) based on unit and individual proficiency at training events. Tracking for these events, as detailed in Navy and Marine Corps Directive

(NAVMC) 3500.65 (Department of the Navy, 2009), is based on the evaluations of leaders, without any specific metrics detailed for success or failure.

The Marine Corps treats LREC skills as enablers to completing tasks. It is currently trying to establish what parts of LREC are necessary to facilitate specific mission accomplishments.

The Marine Corps Regional, Culture, and Language Familiarization (RCLF) Program's curriculum is based on the plan outlined in *Marine Corps Language, Regional, and Culture Strategy* (Headquarters, U.S. Marine Corps, 2011). The program incorporates culture and language training over the course of a marine's career. Different training is provided depending on the mission, role, and rank of the individual. Each marine is assigned a geographical region on which he or she focuses throughout his or her career, with the goal of building a strong and continuing capability within the Marine Corps. Completion of RCLF blocks of instruction will be tracked through the Marine Corps Training Information Management System and Marine Corps Total Force System and coordinated with other training entities within the Marine Corps.

Navy

According to the January 2008 *U.S. Navy Language Skills, Regional Expertise, and Cultural Awareness Strategy* (Chief of Naval Operations, 2008), in the 21st century, LREC capabilities "are essential elements in the Navy's engagement in every phase of war, but paramount to the Navy's ability to shape and influence blue, brown, and green water security environments in all Phase 0 operations."[1] The desired end state includes a "total force that appreciates and respects cultural differences, and recognizes the risks and consequences of inappropriate, even if unintended, behavior in foreign interactions," language career professionals, "other language-skilled Sailors and civilians with sufficient proficiency to interact with foreign nationals at the working level," and "a reserve capacity of organic foreign language skill and cultural expertise that can be called upon for contingencies." During a briefing to the U.S. House of Representatives Committee on Armed Services Subcommittee on Oversight and Investigations, the Navy expressed this as "Language proficiency for some (not all) Sailors. Regional Expertise for some (not all) Sailors. Cultural awareness for all Sailors" (U.S. House of Representatives Committee on Armed Services Subcommittee on Oversight and Investigations, 2008, pp. 40–41).

Other Guidance

The Office of the Under Secretary of Defense for Personnel and Readiness (OUSD[P&R]) is responsible for defense language program policy guidance. The services are responsible for training their forces with appropriate LREC capabilities. However, the combatant commanders are responsible for identifying LREC requirements in support of operations in their areas of responsibility (AORs). Commanders may also publish guidance specifying training that is mandatory for troops deploying to their AORs.

One of the few documents detailing specific language requirements for GPF is the Commander, International Security Assistance Force/Commander, U.S. Forces–Afghanistan (COMISAF/USFOR-A) COIN training guidance first issued by GEN Stanley McChrystal on November 10, 2009 (Headquarters, U.S. Forces–Afghanistan and International Security Assistance Force, 2009). In addition to guidance on basic elements of behavior, this memoran-

---

[1] *Blue* here refers to deep or open water. *Green* refers to coastal water, ports, and harbors. *Brown* refers to navigable rivers and estuaries.

dum states that "each platoon, or like-sized organization, that will have regular contact with the population should have at least one leader [who] speaks Dari at least the 0+ level [sic], with a goal of level 1 in oral communication." This document addresses the training that personnel must receive prior to deployment in theater. Equally important is the fact that DoD is using the CBRIP to identify all COCOM language capability requirements. Results are tracked and endorsed by the Joint Requirements Oversight Council so the LREC demand signals will reach the force providers. These requirements link Universal Joint Tasks to manning documents, establishing the formal LREC requirements for the GPF.

However, as the recent U.S. Government Accountability Office (GAO) report *Military Training: Actions Needed to Improve Planning and Coordination of Army and Marine Corps Language and Culture Training* (GAO, 2011a) presented in detail, there have been differences in the predeployment language and training requirements issued by CENTCOM, U.S. Forces–Afghanistan (USFOR-A), the Marine Corps, the Army, and the Office of the Secretary of Defense (OSD) for GPF deploying to CENTCOM's AOR between June 2008 and March 2011. For example, different sets of guidance require familiarization with phrases in "Afghan languages," "Dari," "Pashto," "Dari and Pashto," or "the language of the region to which they will be assigned." Changing, inconsistent, or ambiguous requirements of this nature make it difficult for individual units and the services to prepare, implement, manage, and track adequate training programs. Note that one of the challenges inherent to the Afghanistan problem set is that the required languages differ from region to region within Afghanistan.

Summary of Policies and Directives

Our review of LREC policy and directives leads to three important conclusions. First, there is a clear belief that LREC skills are important for military effectiveness, especially for COIN. Second, although DoD places equal emphasis on LREC skills, the services emphasize cultural skills over language and regional expertise for GPF. Third, existing LREC guidance appears to be based on anecdotes from the field that have not been collected in any systematic manner.

Review and Analysis of Interviews

One of the goals of the qualitative interviews was to determine how the services and COCOMs view LREC training and capabilities and how they may be using data to assess the linkage between LREC and mission success. The research team used the interviews to verify whether, in fact, LREC practitioners' and policymakers' actions reflect the value placed on LREC according to policy and directives. The information gathered from the interviews also provides an opportunity to look for best practices in terms of data (and data collection) that can be used to empirically support the link between LREC training and skills and mission success.

In this section, we also review relevant literature from business and industry to supplement the interviews. This literature can provide insights into how to assess the link between LREC training and skills and mission success. We then utilize data from the SOF-A to empirically assess the association between LREC training and job performance, one aspect of mission success.

LREC Capabilities and Overall Unit Readiness

All of the individuals interviewed agreed that LREC capabilities were important for unit readiness and mission accomplishment for at least some types of operations. For example, the Marine Corps Center for Advanced Operational Culture Learning (CAOCL) is currently evaluating a 2,500-person survey of redeploying marines. According to one interviewee, the majority of respondents indicated that language and cultural capabilities were "extremely relevant" and that language was "important" to the success of their mission. This was one of the rare instances in which survey data were used to support the importance of LREC. Overall, the majority of the evidence to support this claim was anecdotal.

One interviewee (from an oversight organization) noted that, "in asymmetric warfare of any kind, language capabilities are required."[2] Another interviewee stated that language and culture training and capabilities are needed "across the board," although the exact types and mixes of skills needed depend largely on the operation and the role of the individual service member. The interviewee stated that training is also needed throughout the career of a service member, with different training at different points along the military career. Keeping with the notion of specialized, situation-specific training, one interviewee commented,

> The skills differ in scope: Some need more, and some need less. Some units need more culture and less language or vice versa. We recognize that there is a different mix, balance, or blend depending on your MOS [military occupational specialty] or your rank. One program may need a focused set of skills; a different program may need a more diffuse set. Predeployment infantry may have one set of needs, but if you're going downrange to train foreign forces, you will need a different set. At the service level, there are so many variations for all of this, and we are trying to find a common denominator. Even culture-general and culture-specific have variations depending on the needs of the receivers.

On some occasions, interviewees likened the current conflicts in Iraq and Afghanistan to that in Vietnam, in which interaction with the local populace was of key importance in determining mission success. Another interviewee stated that having service members with some LREC skills "open[s] doors" and dispels myths that "ugly Americans don't even try." Yet another interviewee compared language skills with a "fluffy towel" versus a "thin one." In previous years, language was a "nice to have," or a "fluffy towel." Now it is seen as more of a necessity, or a "thin towel."

This individual also pointed out that, for GPF, it is not essential to have a high level of skill in order to make a difference in terms of mission. If an individual needs to have a basic conversation with a villager, a college-level understanding of language may not be necessary. One interviewee cautioned that, although LREC skills are important for readiness and mission success, "Mission readiness [is] *minimal* requirements [and] standards." This issue of minimum requirements has important implications for tracking LREC skills, especially if minimal skills are defined as basic or low level. For example, if one considers a very low level of language (e.g., 0 or 0+) to be a minimum standard for unit readiness, there must be a way to track who has met it. As will be discussed in Chapter Four, the ability to effectively track very low-range (VLR) language skills (that is, low scores) does not currently exist, although it can be tested in some languages.

---

2   Note that the opinion expressed in this quote did not come from one of the COCOM- or service-specific interviews.

However, not all interviewees indicated that LREC skills are equally valued by all military personnel:

> Language [instruction] is hard for people to consume. They don't think it's important. It's a daily struggle to communicate language importance to commanders. . . . Everyone thinks it's as simple as [using] Rosetta Stone [language-learning software], but it's more than that—it's strategic in nature.

Similarly, several interviewees noted that language in particular is not currently built into units because COIN and nation building—activities that rely heavily on the ability to communicate with nationals—are not the units' primary missions. These interviewees observed that, because service members are trained to fight wars, when a switch occurs from fighting to working with in-country nationals or coalition partners, only then are LREC skills needed within a unit.

Further, the importance of LREC skills may vary throughout different phases of an operation (e.g., shaping operations versus combat operations). Another interviewee pointed out that the real question is the relative importance of these skills in the overall set of skills that a unit needs. Language and culture may be important, but it may be more important for a soldier to be able to know how to use a weapon or tool.

The consensus among the interviewees who shared this view was that the mission would be accomplished with or without language skills, although LREC skills would likely lead to more–efficiently conducted military activities. One interviewee stated,

> Military mindset is to accomplish the mission, no matter how bloody. It might be ugly, but we'll get it done. . . . It's probably true that, if we don't speak the language, we'll take higher casualties and it will take longer, but we'll still win.

Similarly, some interviewees believed that LREC skills would enhance service members' abilities to do their jobs: "Bottom line, why are we teaching the skill? To be better, more-effective airmen, soldiers, and sailors."

Some of these statements indicate that certain aspects of LREC skills may be valued over others, especially for GPF. At least one interviewee noted that speaking a language was more important for GPF than listening skills. Yet other interviewees suggested the opposite— namely, that listening was more important than speaking. Whether or not reading is a required skill for GPF was also not a universally shared opinion among interviewees. Despite inconsistency in what aspect of language was most valuable to GPF, most individuals interviewed focused specifically on the language aspect of LREC skills, although a few dissenting voices were heard. One interviewee stated that, "As a commander, I'd spend more emphasis on culture, and I think I can get that trained to GPF faster than language." And finally, one interviewee suggested that both language and culture are important:

> You have to do both. Culture is limited without the ability to communicate. Language is to communicate intent and be an active listener. Culture is to make the leader more flexible in his [or her] thinking, so people don't come to COIN or security force assistance with an American mindset, which can be very limited. They're both essential.

When asked about the importance of LREC capabilities and training for readiness, interviewees identified three other important themes. First, despite the overall belief among the

U.S. armed forces that LREC skills are important, adequate incentives to not only learn but also maintain skills are not currently in place for low levels of proficiency. Second, service members may be willing to learn a new skill with respect to LREC, but, if those individuals feel that their skills will not be valued or utilized, they may be reluctant to follow through with training. Third, top-level leadership must buy into the importance of LREC capabilities to ensure that their importance for readiness and success are ingrained into the service culture. One interviewee recently attended a conference on LREC and stated,

> [A]rguments put forward by retired generals at this conference [are] that we need to worry about this as much as loading our Howitzers. . . . [We] have to spend more emphasis on culture and language areas as we get ready to go somewhere.

Type of Mission and Capabilities for Language, Regional Expertise, and Culture

The preceding section focused on the importance of LREC training, skills, and capabilities for mission readiness and mission success, according to the interviewees. However, the interviewees made clear that LREC capabilities were not a panacea for success. Instead, they noted that LREC skills are mission-specific enhancers that may be beneficial, or even critical, to success for some missions but not necessary for others. As one interviewee noted, "Readiness-wise, it doesn't play a big role unless you're talking about a specific mission or deployment." On the other hand, another interviewee stated that LREC training skills and capabilities were needed across the board but with a different mix depending on mission, role, and rank.

Anecdotal evidence from interviews suggested some domains and specific types of missions for which LREC capabilities may be most effective. In general, these activities involved interaction with civilians and included COIN operations, humanitarian missions, disaster relief, building partner capacity, counterdrug activities, and security force assistance. One interviewee noted that the importance of LREC skills, and language in particular, for non-kinetic operations is supported by incidents reported in lessons-learned documents from such military efforts as Joint Task Force–Haiti (see, for example, Keen et al., 2010). Perceptions of the importance of LREC skills also vary by branch of service; for example, a Navy interviewee stated, "You don't pull into port without engaging. We are conscious that we are a diplomatic tool."

Recognizing that LREC skills are mission-specific, the Director's Action Group of the JCS Manpower and Personnel Directorate (J1) is currently in the process of standardizing assessments of LREC skills necessary for geographic COCOM missions. The endeavor uses a standard methodology for determining baseline LREC capabilities for military tasks taken from the Universal Joint Task List (UJTL). According to these requirements, which will ultimately serve as a demand signal, the services will be able to generate forces with the required capabilities. This coordination effort may alleviate some of the tension between combatant commanders and the services that was described during the interviews: COCOMs may not always know what they need, and the services are not always sure whom, and what, to train.

These newly established LREC standards will eventually be tied to METLs in order to facilitate planning on a more tactical level. Through conversation with geographic COCOM senior language authorities, COCOM operations directorate (J-3) and plans directorate (J-5), and other relevant individuals and offices, the JCS research team is determining what skill (e.g., language, regional expertise, or cultural knowledge), at what level, and how many individuals are needed for each specific task. Thus, as the analysis process continues, it may become evident that, for some tasks, GPF do not need to possess any LREC skills, whereas, for others, such skills may be deemed mission essential.

The linkage of LREC capability requirements to tasks and missions is vital information for decisionmakers. For that reason, the CBRIP requires every COCOM to specify as it develops its LREC requirements the importance of desired LREC capability to the accomplishment to the Universal Joint Task and of the Universal Joint Task to the accomplishment of the mission." This aids in the prioritization of requirements within the COCOM and among the COCOMs.

Despite the progress on determining LREC requirements, much work remains. As one interviewee stated, "Even when done with the JCS methodology, until you can validate it more, it's a guess." Another noted that there was "no research linking language skill to mission readiness or accomplishment," which became a recurrent theme in the interviews.

Determining the Importance of LREC Training and Capabilities

Although evidence linking LREC training and capabilities to mission readiness and mission accomplishment is sparse, many interviewees offered anecdotal evidence to support the hypothesis that they are linked. Several interviewees noted that previous conflicts (e.g., Vietnam, the Balkans) should be viewed as "lessons learned" in terms of the importance of LREC skills. At the same time, however, they noted that those lessons learned would "never be able to tell us what problems with local people were avoided because of using the appropriate behavior for the situation." In this sense, the counterfactual (i.e., what would have happened had service members not used the appropriate social and cultural skills in a certain situation) cannot be known.

Another interviewee noted that, in his branch of service, service members train for convoy operations in a particular region of the world using role-playing activities in which observer-controllers rate the service members on their performances. Trainees are not explicitly rated for their use of LREC capabilities but do receive a go/no-go rating after completion of the training. Thus, one could conclude that raters are *implicitly* factoring LREC skills into these ratings, suggesting they are important for accomplishing the mission.

Other interviewees offered suggestions for how to link LREC training and capabilities to mission readiness and success, including means of data collection. They suggested tying tests of LREC training to objectives. For example, to assess effectiveness of LREC training for mission accomplishment, performance should be tested in a behavioral training exercise (as mentioned in the preceding paragraph). They also suggested collecting real-time feedback from commanders on the ground. One interviewee provided an example from Haiti, following the recent earthquake, where commanders were able to use instant feedback, via daily briefs, to adjust LREC skills for the humanitarian mission (e.g., if language skills in a particular dialect were needed, someone with that skill could be found). Interviewees also suggested conducting systematic interviews with service members who have just returned from theater to provide evidence that could be used to support an LREC–mission linkage.

One interviewee offered,

> Talk to commanders, especially multiple deployers, and ask them the value [of LREC skills.] I know it's anecdotal, but the sound bites and comments—good, bad, or indifferent— you'll get will be the most valuable thing you can provide for measures of effectiveness.

One interviewee called this method a "love meter" but said that it was the best that could be done.

Interviewees also pointed out that service members may inadvertently provide anecdotes that LREC skills are tied to mission readiness and mission accomplishment through blogs and memoirs or via social media, such as Facebook or Twitter. Researchers could mine such resources with social-media tools. Note that public release of such information may present a security issue.

Review of the Literature on Language and Culture Training in Business and Industry

In thinking about how to go about testing whether or not LREC training or skills have any association with mission objectives, it may be helpful to look to two separate but related literatures, both from industrial and organizational psychology. The first literature focuses on the importance of language and cultural knowledge among expatriates in business and industry, and the second focuses on training effectiveness in general.

A key issue for multinational corporations is the ability of expatriate managers to effectively adjust to a host country's social and business environment. When managers fail to successfully navigate this transition, it can be very costly to corporations,[3] and a large literature examines what role language skills and cultural expertise play in predicting manager success. In general, this literature finds that some knowledge of the host nation's language and culture is helpful for manager success in terms of adjusting to (1) work roles, (2) interaction with host nationals, and (3) the general culture and everyday life (see Black, 1988).[4] For example, Brislin and Yoshida's (1994) review of the literature shows that the benefits of what they call cross-cultural training (CCT) fall into three categories: positive effects on people's thinking and knowledge development about other cultures; positive effects on people's affective responses (e.g., attitudes, self-concepts, emotions, and feelings of comfort with another culture); and positive effects on people's behavior when interacting within another culture (see also Black and Mendenhall, 1990; Harris and Moran, 1996; Kealey and Protheroe 1996; Mendenhall et al., 2004). Language skills, important in their own right, play a large role in facilitating cross-cultural competency (Brewster, 1995; Puck, Kittler, and Wright, 2008).

These examples suggest that predeparture CCT (i.e., training that is done prior to the person's move to a foreign country) should be associated with a successful transition into the host country; however, the evidence is mixed (see Puck, Kittler, and Wright, 2008). Some studies have reported a positive association between predeparture CCT and expatriate adjust-

---

[3]   A study by Vogel, Van Vuuren, and Millard (2008) estimates that the cost of a single failed assignment in a multinational corporation is between $40,000 and $1 million (in U.S. dollars).

[4]   There is no universal agreement in the expatriate literature on what it means to make a "successful transition." As noted earlier, measures of *adjustment* generally involve assessing an individual's ability to adjust to work roles, interacting with host nationals, and the general culture and everyday life (see Black, 1988), although adjustment may not necessarily correlate with *success*. A recent analysis suggests that there may be as many as nine different aspects of expatriate *success*, including cultural adjustment, work-related adjustment, career development, headquarters (HQ)–subsidiary coordination, assignment completion, profession and skill development, shaping and controlling the subsidiary, satisfaction, and overall assignment effectiveness (Hemmasi, Downes, and Varner, 2010). And as Abbe, Gulick, and Herman (2007) notes, cross-cultural success in the military includes the added dimension of being able to exert influence *across* cultures.

ment (see Black and Mendenhall, 1990; Deshpande and Viswesvaran, 1992; Mendenhall et al., 2004), whereas other studies have found no association, or even a negative one (Black and Gregersen, 1991; Gregersen and Black, 1992; Kealey and Protheroe, 1996; Puck, Kittler, and Wright, 2008). Two possible reasons for this inconsistency are the variability found across training programs (e.g., in length, content, and degree of specificity) and differences in how *success* was defined.

For example, Black and Mendenhall (1990) identifies two different types of CCT: information-giving and experiential learning.[5] Under information-giving, the authors identified practical information, area studies, and cultural awareness as distinct techniques for training programs. Under experiential learning, the authors identified only intercultural effectiveness skills as an approach, resulting in a total of four different training methods. Gudykunst, Guzley, and Hammer (1995) added the distinction between culture-general and culture-specific training in their overview of different extant cultural training techniques (see also Abbe, 2008; Abbe, Gulick, and Herman, 2007). Kealey and Protheroe (1996) identifies four different approaches to CCT: practical information, area studies, cultural awareness, and intercultural effectiveness skills. However, the authors note that the empirical evidence as to the effectiveness of these different types of training is largely inadequate. Thus, the second important literature to examine when considering the impact of LREC training on an outcome of interest (e.g., mission readiness and completion) involves evaluation of training.

Kirkpatrick (1959a, 1959b, 1960a, 1960b) is often credited with developing the first framework for training criteria, as well as evaluation of training. His framework is still among the most widely used today. By far, the most common method of evaluating training is via trainee reactions (e.g., how much the person enjoyed the training, whether he or she felt that it was helpful); however, these types of measures may or may not be related to more-meaningful evaluation data and methods. Kirkpatrick developed a four-level taxonomy that characterizes training criteria. These four criteria can be thought of as measures of training effectiveness (see Alliger and Janak, 1989). The first is a trainee's *reactions* to the training, including subjective measuring of liking or utility. The second is *learning*, or the extent to which the trainee absorbed principles, facts, techniques, or other content. The third is *behavior*, which is related to whether or not the trainee uses techniques and skills learned during training or changes behavior because of training. The fourth is *results*, which Kirkpatrick broadly defined as the desired end product of training (e.g., reduced absenteeism, reduced turnover, increased morale, increased productivity).[6]

Although simple and widely used, the Kirkpatrick model of training evaluation and effectiveness does have limitations (Alliger and Janak, 1989; Alliger et al., 1997; Alvarez, Salas, and Garofano, 2004). The four criteria were originally labeled *steps*, indicating that they result from a linear progression of reactions to learning to behavior to results. In fact, it is more likely that the steps form a feedback loop in which learning affects behavior but behavior and results reinforce one another. A critique that has been leveled against Kirkpatrick's taxonomy is that it

---

[5] This dichotomous approach to forms of training is used by other authors, albeit with slightly different wording. For example, Gudykunst, Guzley, and Hammer (1995) use the term *didactic* in the place of *information-giving* but assign this term essentially the same definition.

[6] Beginning with Becker's (1964) seminal work, there is a large literature in economics on the returns to human-capital investments (e.g., the effect of training on workers' future productivity measured as earnings). This work most closely resembles the *results* training criteria in Kirkpatrick's taxonomy.

assumes that there is a causal relationship between training and the four steps, when, in reality, this is a hypothesis that must be tested. As a result of this critique, Alliger and colleagues (1997) proposed an augmented version of Kirkpatrick's original model.

The updated taxonomy separates reactions into affective reactions to training (e.g., "I liked the training") and utility judgments (i.e., that the training has practical value). It measures learning as immediate posttraining knowledge, knowledge retention, and behavioral or skill demonstration. These three subcategories allow for the assessment of training effectiveness over time. The updated taxonomy replaces behavior with *transfer*, which refers to on-the-job usage of new skills. Although behavioral and skill demonstration may occur at any point in any setting, transfer is distinguished by its setting. Finally, the updated Alliger et al. model retains Kirkpatrick's original definition of *results* but notes that they are often the most difficult to obtain.

Despite some differences between the expatriate and military experiences, according to the training effectiveness literature,[7] it is clear that a well-rounded assessment of LREC training should include elements of all four evaluation criteria—reactions, learning, transfer, and results—even if the actual LREC skills needed by service members differ from those of civilians in business and industry (see Abbe, Gulick, and Herman, 2007). Trainee reactions to training are the easiest to collect, and, according to the interviews in this study, this is a typical evaluation technique already in place. DLIFLC has an evaluation unit, but, because the training it provides is so varied and tailored to the needs of the language user (i.e., the services), there is no universal or systematic evaluation of training.

Some of the DLIFLC evaluation procedures include pre-, during- and posttraining interviews with students. They also include field surveys of service members who are mid-tour, and after-action reports (AARs) from service members returning from deployment. However, different elements of these data could potentially be used to assess all four of Kirkpatrick's evaluation criteria. The Defense Language Proficiency Test (DLPT) can also be considered a measure of learning because DLPT scores after training represent knowledge retention, as defined by Alliger and colleagues (1997).

A recent two-phase study by the Cognitive Performance Group (2010, 2011) used "K1" (reactions) and "K2" (learning) approaches to assess predeployment cultural knowledge and survival language skills across the four DoD services. Through collection of language and culture training materials; interviews with providers, trainers, and students; and survey data, the research team found that culture knowledge training received higher ratings for satisfaction, usefulness, and relevancy than did survival language training. The research team also noted that not all services utilize methods of training evaluation that include pre- and posttraining measures, making it difficult to assess whether or not trainees' levels of knowledge increased during the training period.

From the interviews, the research team identified few assessments of learning and transfer and no assessments of results. One possible reason for the lack of studies with respect to assessing results of training, or what could be thought of as the desired end product of LREC

---

[7] One important difference between the expatriate and the military experience is that service members typically do not have the option to terminate their stay in a country. As Abbe, Gulick, and Herman (2007) notes, the implications of this difference are far reaching. If, for example, service members (who are unlikely to have the option of leaving their assignment) have inadequate CCT or competencies, they may leave long-lasting impressions on the people with whom they engage. Ultimately, these impressions may hinder the effectiveness of future missions in that area of operation.

training, is a lack of established measures. COL Ralph Baker (2010), in a commentary about his experience as a brigade combat team commander in Iraq, offered suggestions for measuring mission success (i.e., results),[8] such as number and categories of people waving as a patrol moves through a village, the amount and type of graffiti on village walls, and the amount of negative press in local media (see also Murphy, 2009–2010). With the help of social scientists, commanders on the ground could compile a list of similar metrics that could be used to link the LREC training and skills of a combat team to mission success. Murphy (2009–2010), however, cautioned that such methods of assessing the effectiveness of LREC training, or of strategic communication, are unlike evaluations of kinetic missions, in which feedback is almost instantaneous. There may be significant time lags between the behavior learned from the training and the resulting change in local attitude. Moreover, many other factors may influence local attitude and its expression. Thus, it is important to think about what types of measures of training evaluation and success would be most meaningful in this context.

Analysis of the Status-of-Forces Survey

Because the literature review and interviews provided little systematic evidence of an empirical linkage between LREC training and capabilities and military mission readiness or mission effectiveness, we also searched for quantitative data that would allow us to estimate this association. We found no existing data sets that would allow for such an analysis. However, as noted in Chapter Two, the June 2010 SOF-A does contain a small selection of questions about LREC training.

The analysis that follows focuses on a very specific group of service members.[9] Because the LREC-related questions in the SOF-A are about predeployment training, the research team focused on service members who had experienced a deployment of at least 30 days in the preceding two years (as of the date of the survey). Further, they restricted their analysis sample to those service members who also experienced at least one deployment in OIF or OEF. As shown in Table 3.1, roughly 44 percent of all recent deployers had also experienced one deployment in OIF or OEF since 9/11.[10] However, it is important to note that the data did not allow the team to determine whether the *most recent* deployment was to Iraq or Afghanistan.

Table 3.2 shows the percentage of recent deployers with a history of deployment in OIF or OEF who self-reported that they received any predeployment language, region, or culture

---

[8]   It should be noted that Baker's observations are based on his use of information operations, which are related to but not synonymous with LREC skills.

[9]   Results were similar across all services with three exceptions. First, the Army and the Marine Corps had a larger percentage of service members who reported receiving language, regional, or culture training. Second, Air Force service members reported using more distributed learning and online training than their peers in the other services. And third, more Navy and Marine Corps respondents indicated that their LREC training took place at an "other" location (i.e., not in a classroom, training exercise, or home station or through distributed or online learning).

[10]   CIs for all estimates are presented in Appendix D. Because the SOF-A data use sampling statistics to select prospective respondents rather than a census framework, in which all possible respondents are interviewed, there is some degree of error in estimating means, percentages, and other statistics. CIs represent the range of sample statistics (e.g., percentages) expected given the sampling strategy used. The 95-percent CI reflects the range in which 95 percent of those sample statistics would fall.

Table 3.1
Deployment Characteristics (%)

Deployed in the Two Years Prior to the Survey and with at Least One Deployment in OIF or OEF	Total Force	Officers	Enlisted
Yes	43.7	55.6	41.3
No	56.4	44.4	58.7

SOURCE: DMDC, 2010a.

Table 3.2
Type of Predeployment Training Received Among Those Deployed in the Two Years Prior to the Survey and for One Deployment in Operation Iraqi Freedom or Operation Enduring Freedom (self-reported)

Training Received	Total Force	Officers	Enlisted
Language	27.6	23.3	29.0
Region	50.9	51.4	50.8
Culture	59.9	62.5	59.2
Any LREC	61.5	64.0	60.8

SOURCE: DMDC, 2010a.

training.[11] Approximately 28 percent of all recent deployers reported receiving predeployment language training, 51 percent reported receiving regional training, and 60 percent reported receiving culture training. Overall, 62 percent reported receiving at least one type of predeployment LREC training. These percentages are similar across officers and enlisted personnel. Subsequent results focus on those recent deployers who reported having received any language, region, or culture training. An important caveat is that receipt of training is self-reported. Thus, some individuals who did receive training may not have perceived it as such and therefore may not have acknowledged LREC training receipt in the survey.

The SOF-A data were collected between June 11, 2010, and July 22, 2010, and focused only on active-duty members. Of the 68,540 active-duty service members eligible to participate, 18,983 returned usable surveys, resulting in an adjusted weighted response rate of 25.3 percent, with the results weighted to reflect overall population values, presented for all service members.

Figure 3.1 presents the duration of training in hours for those who reported receiving any LREC training. The majority of service members received predeployment LREC training that was less than a full day, or eight hours, in length. Only 20 percent received training that lasted more than nine hours, and the smallest percentage, approximately 5 percent, received training of 40 or more hours.

Figure 3.2 shows the distribution of recent deployers with any LREC training by the type of trainer (i.e., Service Culture Center, DLIFLC, contracted subject-matter experts, or trained unit trainers). Respondents were allowed to select more than one type if more than one individual or group was responsible for the training. The majority of trainers were from the service

---

[11] Note that the series of questions in the SOF-A asks about predeployment LREC training that occurred before a deployment to Iraq or Afghanistan, although it may not be the service members' most recent deployment.

Figure 3.1
Distribution of Predeployment LREC Training, by Duration (in hours) Among Recent Deployers

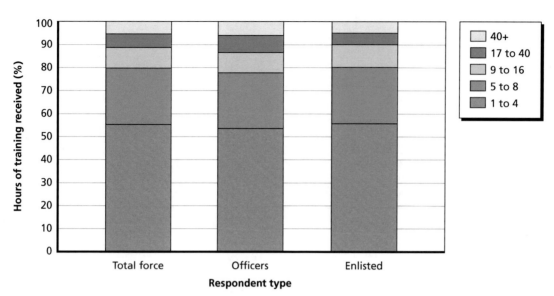

NOTE: The figure applies to those who had deployed at least once in the two years prior to the survey and had been deployed in OIF or OEF.

RAND TR1192-3.1

Figure 3.2
Predeployment LREC Training, by Type of Trainer

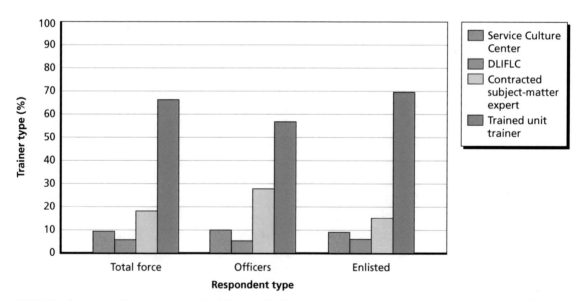

NOTE: The figure applies to those who had deployed at least once in the two years prior to the survey and had been deployed in OIF or OEF at least once since 9/11.

RAND TR1192-3.2

member's own unit, followed by contracted subject-matter experts, Service Culture Centers, and DLIFLC.

In terms of where predeployment LREC training took place, most occurred within the classroom setting (55.2 percent). This was followed by training that occurred at the service member's home station (24.5 percent), another unspecified location (13.5 percent), distributed learning or online (6.8 percent), and in the context of a training exercise (5.9 percent). This distribution did not differ significantly by officer or enlisted status. More than half of all service members (55.8 percent—54.5 percent of officers and 56.2 percent of enlisted) indicated that the predeployment LREC training they received included some type of supplemental take-away material. The majority, roughly 86 percent, received a pocket handbook or smart-card reference aid. Another 10 percent received a CD or DVD, and another 4 percent received some other type of supplemental aid.

The SOF-A also asked respondents to indicate which objectives were included in the predeployment LREC training they received.[12] Table 3.3 presents the percentage of recent deployers who indicated that a specific objective was present in their predeployment training. With the exception of the economy and negotiations, more than two-thirds of the training the recent deployers received covered all of the other possible training objectives. The most frequently covered objective was religion (with more than 90 percent coverage). Geography, social organizations, cross-cultural competency, history, and political structure were also common objectives. When asked whether the training met specified objectives, 89 percent of all service members (86.4 percent of officers and 89.1 percent of enlisted) indicated that it did.

Respondents reported mixed satisfaction with predeployment LREC training. Figure 3.3 shows the percentage of recent deployers (who self-reported receiving any type of LREC training) who were very satisfied, satisfied, neither satisfied nor dissatisfied, dissatisfied, or very dissatisfied with the training they received. Keep in mind that the report does not indicate whether the respondent was rating the training before or after a deployment. Roughly 40 per-

Table 3.3
Respondent-Reported Meeting of Predeployment LREC Training Objectives (%)

Objective	Total Force	Officers	Enlisted
Geography	87.7	88.8	87.4
History	79.3	82.9	78.2
Cross-cultural competency overview	81.1	78.1	82.0
Political structure	77.0	74.5	77.8
Social organizations	83.2	79.3	84.4
Religion	93.6	96.1	92.8
Economy	72.9	69.3	74.0
Negotiations	51.6	47.5	52.9

SOURCE: DMDC, 2010a.

NOTE: The table applies to those who had deployed at least once in the two years prior to the survey and had been deployed in OIF or OEF at least once.

---

[12] It could not be verified whether respondent perceptions coincided with the objectives as conceived of by the developers of the training.

Figure 3.3
Satisfaction with Predeployment LREC Training

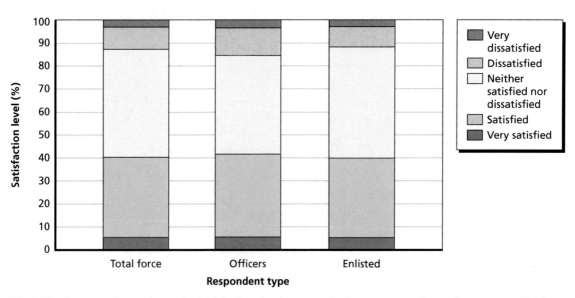

NOTE: The figure applies to those who had deployed at least once in the two years prior to the survey and had been deployed in OIF or OEF at least once since 9/11.
RAND *TR1192-3.3*

cent of all service members were either very satisfied or satisfied with predeployment LREC training. However, a larger percentage—roughly 47 percent—were essentially neutral with respect to satisfaction. Finally, approximately 13 percent were either dissatisfied or very dissatisfied with the training they received. Results were consistent for both officers and enlisted. The survey data do not explain *why* service members may have been dissatisfied with the LREC training they received.

Finally, the SOF-A asked whether or not predeployment LREC training was beneficial in performing the respondent's job (see Figure 3.4). Although 40 percent of respondents were satisfied or very satisfied with the training they received, only 36 percent (33.0 percent of officers and 37.6 percent of enlisted) indicated that the LREC training they received was beneficial to performing their jobs. Unfortunately, the SOF-A does not indicate the service members' jobs, whether they were attached to language-specific billets, or whether they were members of the GPF. Further, it is not clear how job performance is tied to mission readiness or accomplishment. One could assume that anything related to job performance would aid in the successful completion of mission objectives.

One existing study does, in fact, suggest that specific LREC skills may be useful for mission accomplishment because of their linkage to the ability to do a job. Hardison and colleagues (2009) identified 14 potentially important cross-cultural behavior categories from the academic literature. With the help of Air Force subject-matter experts, the authors grouped them into enabling (i.e., skills typically used to facilitate day-to-day activities and likely to be used across a variety of jobs) and goal-oriented behaviors (i.e., skills typically associated with mission-specific activities and likely to be used only by selected Air Force specialty codes). The researchers then surveyed recently deployed airmen to determine the importance of the categories for different deployed Air Force jobs. Following is a list of enabling behaviors, followed in parentheses by the items' order of importance, as rated by the airmen in the survey:

Figure 3.4
Respondents Reporting That LREC Training Was Beneficial to Job Performance

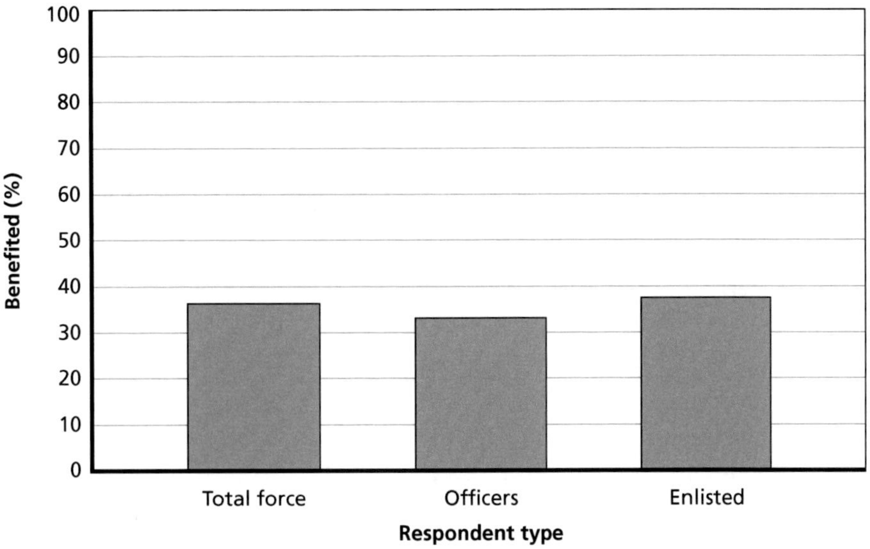

NOTE: The figure applies to those who had deployed at least once in the two years prior to the survey and had been deployed in OIF or OEF at least once since 9/11.
RAND *TR1192-3.4*

- gathering and interpreting observed information (1)
- respecting cultural differences (2)
- applying appropriate social etiquette (3)
- changing behavior to fit the cultural context (5)
- verbal and nonverbal communication (6)
- managing stress in an unfamiliar cultural setting (7)
- applying regional knowledge (9)
- self-initiated learning (11)
- foreign-language skills (12).

Following is a list of goal-oriented behaviors, also showing in parentheses the airmen's ranking for importance:

- establishing credibility, trust, and respect (4)
- negotiating with others (8)
- resolving conflict (10)
- influencing others (13)
- establishing authority (14).

The highest-rated cross-cultural skills were gathering and interpreting observed information, respecting cultural differences, and applying appropriate social etiquette, which are *general* capabilities that can benefit service members across different job types. In contrast, more-specific skills (e.g., influencing others, establishing authority), which may apply to only a selected group of jobs, were rated least important. Given the wide emphasis on language skills in many quarters, it is interesting that foreign-language skills were rated near the bottom of the 14 categories. Importance ratings of the 14 cross-cultural skills also significantly varied by Air Force specialty code, with those in security forces, contracting, special investigations, and international affairs rating many of the categories as high in importance. In contrast, pilots did not consider many of the skills of "even mild importance" to their jobs.

At least two other studies suggest that general, low-level language, cultural, and regional skills may be useful for job performance. In the first, the U.S. Army Research Institute (ARI) conducted a survey of returning military advisers from Iraq and Afghanistan, interviewing them about their experiences and training (Zbylut et al., 2009). Advisers did not consider language proficiency to be necessary for successfully performing their jobs. However, knowing common words and phrases, local norms, and foreign counterparts' worldviews were rated as necessary skills and seen as important for performance.

In the second study, the Marine Corps, through CAOCL, gathered survey data from just over 2,000 marines, across all occupations and ranks (except flag officers), the vast majority of whom had been recently deployed (CAOCL, 2011a). Among marines who had received predeployment language or culture training, 71 percent stated that their language training made them more operationally effective. Roughly 80 percent stated that their culture training had made them more operationally effective. The difference between the percentages was statistically significant.

In a subsample of respondents with ground combat–related occupational specialties, those marines in ground combat arms both reported using language and culture skills more often and rated them more highly than their combat support counterparts (CAOCL, 2011b). One reason for the difference may be that combat arms marines indicated they spent more time with locals than did marines in combat support occupations. Both combat and combat support marines said that culture *and* language skills were as important as more-traditional combat training skills.

Taken together, the results of the CAOCL Culture and Language Survey suggest that both language and cultural skills are important for mission success, although culture was rated as more important. However, two caveats of the study are worth noting. First, the sample is not necessarily representative of all marines. The authors note that marines at the lowest ranks (e.g., E1s, E2s, and O1s) are underrepresented. Presumably, many of these marines make up the GPF. Nor are the results weighted to reflect the overall population of marines. Second, the data cannot be used to establish a *causal* link between language or cultural skills and mission success.

Summary of the Status-of-Forces Analysis

From this analysis of the SOF-A data, we found that more than half of all recently deployed service members had some LREC predeployment training, with most of that training coming in the form of cultural or regional knowledge. This training typically lasted for less than a full day and occurred within the service member's own unit. Satisfaction with training was generally positive, although a substantial proportion (roughly 45 percent) had a neutral view of their training. Although the majority indicated that the training met its intended objectives, just over one-third of recent deployers said that the LREC predeployment training they received was beneficial to performing their jobs. However, one existing study (Hardison et al., 2009) suggests that the importance of LREC skills varies by job. If this is the case, our analysis of the SOF-A data would not be able to reflect the differences. Given the limitations of the SOF-A data—namely, that they cannot be used to identify causal relationships because they are self-reported survey data—these data should not be viewed as providing a definitive answer to the question of whether LREC training and skills are associated with mission readiness or accomplishment. The results simply provide a first step toward empirically addressing this issue.

CHAPTER FOUR

The U.S. Department of Defense's Ability to Track LREC Training and Capabilities

Despite an increased recognition of the importance of LREC training and recent efforts to improve that training, there are still difficulties in tracking either the expertise of warfighters or the training they have received. With the exception of language proficiency (as evaluated by the DLPT), there are currently no measurable standards that can be incorporated into any database for tracking skills of individuals or units. This lack of tracking can have serious implications for the services' ability to utilize the right people, with the right skills, in the right times and places. As noted in a recent GAO (2011b, p. 20) report,

> By not capturing information within service-level training and personnel systems on the training that general purpose forces have completed and the language proficiency gained from training, the Army and Marine Corps do not have the information they need to effectively leverage the language and culture knowledge and skills of these forces when making individual assignments and assessing future operational needs.

Further, the ability to track language skills from proficiency tests is also problematic, with noted limitations. Several interviewees pointed out that the standard DLPT does not effectively measure VLRs of language proficiency, does not test for proficiency in all languages of interest to the military, and does not test for high levels of proficiency for some languages. Moreover, although there is cultural content in the new DLPT, the test is not intended to focus on the language and cultural skills needed to complete specific tasks (i.e., it is not task-oriented). As of the date of this report, there were seven VLRs being used to measure low-level language proficiency.

Training delivery is frequently tracked, but there is no central database that tracks the variety of training courses any given warfighter has taken (see also GAO, 2011b). Rather, units track training in predeployment as part of their certification. The instructors provide certification for those who complete training, and the records are typically tracked through the online systems delivering web-based training (although, in some cases, no tracking or course taking or certification is done). As a result of the fact that these tracking systems (e.g., those used by different instructors) are different, it is difficult, if not impossible, for commanders to accurately determine the current training and qualifications of individuals or units.

Most evaluation of readiness is based on performance in field and training exercises. For example, marines train at Mission Mojave Viper; the Army trains soldiers at the National Training Center and Joint Readiness Training Center. The results of these training exercises are qualitatively assessed by evaluators and are discussed with commanders and other leadership. However, interviewees pointed out that there is currently no operationally defined measurement for language and cultural readiness or proficiency.

Currently, the DLO is working to create a combined database for tracking qualifications of individuals, the Language Readiness Index (LRI) module for the DRRS. The DRRS-LRI will provide planners with the DLPT scores of individual warfighters from all the services and allow them to plan for personnel needs appropriately. Although there are some elements of language skill that the DRRS-LRI system will not track (e.g., English-language proficiency, dialects of specific languages), this project is a significant step forward in keeping track of proficiencies for planning purposes.

This chapter discusses interviewees' responses to questions regarding current tracking of LREC skills. These discussions lead to an analysis of how to improve tracking, thus enabling more-efficient planning with respect to existing LREC capabilities of the GPF. Linking training to readiness requires sufficient data on that training to assess its impact, and obtaining such data is not possible without accurate tracking of different courses of instruction and proficiencies. This discussion of tracking is essential to understanding readiness within the services.

Analysis of Interviews

We designed interview protocols to discuss tracking of LREC training and to solicit information on how interviewees felt about the relationship between training and readiness. DoD can determine the effectiveness of LREC training on promoting mission readiness and effectiveness only if the training is properly tracked. To that end, we asked questions about the tracking of both training and capabilities, including how or whether self-reported or heritage skills are tracked. In addition, we asked interviewees about what data they would like to have available in a tracking database to make better decisions regarding planning and assignments.

In response to the JCS project mentioned in Chapter Three, both the services and the COCOMs have begun assembling data on their current training and capabilities. Many interviewees specifically mentioned this data call during interviews and referenced the data call as a first step toward assembling a single database of LREC training and capability across the force. However, current tracking efforts for LREC training across DoD are not standardized, and most data are spread across multiple, disconnected databases. For example, within the Army, training may be captured in the Army Training Requirements and Resources System (ATRRS), Status of Resources and Training System (SORTS), Digital Training Management System (DTMS), or other systems, depending on the level of instruction, whether a unit or an individual went through the training, and the service or organization that provided that training.

For commanders to determine their current capabilities, their staff must assemble data from these disparate sources. As identified by one interviewee, "part of the issue now is that those bits of information exist but not all together. DTMS has DLPT scores, a personnel record has deployment history, but it's not unified." Interviewees from DMDC stated that they pull their information from the same disparate personnel systems as the services and COCOMs. The LRI is an application that draws its data from the DRRS data warehouse, which, in turn, draws data from DMDC and the services. It portrays the language capability of the Total Force in a way that allows customized searches linked to exportable Excel spreadsheets. The DLO is implementing plans to incorporate regional proficiency ratings for personnel.

Aside from the services, individual organizations also have their own systems for tracking training. For example, SOUTHCOM has the SOUTHCOM Enterprise Management System (SCEMS), which includes self-reported fluencies in a variety of languages, based on an intake survey and regular (triennial) surveys of the command. USNORTHCOM also maintains its own learning tracking system. Similarly, DLIFLC maintains records for students who have gone through its Rapport training, and the Navy maintains similar records. However, at the moment, there is no centralized information storage, and the majority of the training that is tracked is only language training. Interviewees expressed the belief that personnel offices within various commands track background knowledge, such as heritage speaking or cultural awareness from family history or deployments, but, as with other tracking, there does not appear to be any standardized and structured database to maintain those data.

According to interviewees, different commands and services determine eligibility for the FLPB in different ways (see also U.S. House of Representatives Committee on Armed Services Subcommittee on Oversight and Investigations, 2010), making it difficult to properly track capabilities using this system. Data collected by the DMDC are gathered from the services and are thus dependent on the guidelines those services use to determine who qualifies as language-capable, rather than on a comprehensive examination of the entire force.

According to interviewees, in some cases, DoD does not properly track resources even when planners are aware of the potential. For example, one interviewee stated that there are "5,000 green card holders in the Army, and we don't track them or what languages they have." In another example, a Marine Corps interviewee described that some groups do not send people to take the DLPT or Oral Proficiency Interview (OPI). The Marine Corps does use VLRs but does not pay FLPB on the basis of VLRs. According to one interviewee, the LREC skill set has to be specific to a task, so the groups focus on specific tasks in their language testing. However, the DLPT tests for global proficiency, such as "how well you can order tea, or listen to the radio, and we don't find that useful." The interviewee continued, "We have to tie this language requirement to tasks, which doesn't test well on the global DLPT."

Most interviewees expressed the desire for a testing program that would be better able to capture lower levels of language ability because the current DLPT is not effective at discriminating language proficiency below a level 2. Although U.S. Special Operations Command (USSOCOM) requested that DLIFLC develop a specific VLR DLPT, it is still being tested and validated and is neither commonly used nor available for all languages (EduMetrica, 2010). As of the beginning of FY 2012, four VLR tests and three interim VLR tests have been fielded to meet the needs of SOCOM and the GPF. The department has the capability to test in a language even if no DLPT does not exist. There are many OPIs available to test at a lower level in many other languages, as well as commercial, off-the-shelf assessment tools. However, other assessment tools should be chosen carefully to ensure the use of valid tests. Although, according to the interviewees, there is strong internal consistency among the different components of DLPT scores (e.g., reading, listening, speaking) at level 2 or above, this consistency does not exist below level 2, making it difficult to operationalize and measure individuals who perform at that level (see also EduMetrica, 2010). General McChrystal's memorandum calling for 0+ speakers among warfighters who will engage the population highlights the importance that some commanders have placed on low-level language ability. Thus, it is essential that there be a reliable and valid method for testing this low level of language ability.

According to interviewees from DMDC, older records were not flagged with whether the recorded level of language skill was self-reported or formally assessed. Although the new tracking system being developed does have a field to track this information, there is no way to determine the difference between the two categories in older records. DMDC interviewees did note, however, that the services can differentiate between self-reported and formally assessed language ability. If an individual is receiving an FLPB, that individual was formally tested.

According to interviewees, tracking of cultural training, which consists primarily of pre-deployment training conducted for deploying units, is minimal. This training is mainly tracked by the services at the unit level and is usually rated on only a go/no-go scale for completion. An interviewee commented that the Marine Corps has a culture of "commander's discretion." He went on to say,

> Our policy is that it is the burden on the commander for determining a requirement. The commander will look at his [or her] mission and develop his [or her] own training plan based on his [or her] own beliefs. The Marine Corps has developed guidance that LREC is important but is up to the commander.

He stated that the main assessment of LREC readiness is made by the commander at a "gut level" based on performance at Advanced Mojave Viper.

One interviewee suggested incorporating training requirements into currently extant tracking systems, such as SORTS: "I think we can drive this with support to levy a T rating [the training status of a unit as identified in SORTS] for a unit commander based on cultural proficiency and language." By incorporating cultural and language proficiency into the overall T rating for units, that proficiency would be specifically tied to unit readiness. However, it might be challenging to design the specific mechanisms to incorporate those proficiencies into the ratings. For instance, the Marine Corps is considering "creating a COIN METL, but there will be so many subtasks that LREC might be part of it, that not getting LREC isn't going to change that red."[1] The interviewee added,

> Unless you elevate LREC to a core competency, it's going to get lost. Since it's only one subtask, it won't be essential. Let's say I have a 1,000-man unit. Let's say 60 people can't fire their weapons. That won't turn that METL red. I don't see how LREC would integrate into some sort of evaluation.

In addition, much cultural training received by a deploying unit is incorporated into other mission training: "They train on convoy [operations] in a particular region, and the observer-controllers rate them but not on the cultural aspect of the task. They get cultural elements in their training, but it isn't rated or recorded." One reason for this is that there are often more training requirements for a deploying unit than can be physically performed by that unit in a set amount of time. For example, one interviewee commented that there are 77.5 days' worth more requirements than there are days to accomplish training, so "where the LREC training falls is up to the commander's priorities."

Frequently, the shortage of training time means that some cultural training must be combined with other training. Because cultural training is often enmeshed with tactical training, it is difficult to track the training received by a unit or individual. Even when the training explic-

---

[1]   In this context, *green* means meeting requirements or ready to deploy. *Red* means not meeting requirements or not ready to deploy.

itly addresses cultural knowledge, this training is stored in a wide variety of formats, including both electronic and paper documentation, and rarely uploaded to any central database.

Given that cultural ability is not defined in operational terms, there is no standardized method to determine the level of cultural readiness any individual or unit may possess, even if all of this training were tracked and maintained in a central database. In addition, without a standardized definition of *regional expertise* and *cultural skill level*, the identifier used by one organization could easily be different from another's, or, as one interviewee questioned, "How do you make it uniform across the services? How do I know that a Navy guy with a 'region 3' rating, that it's the same as an [Air Force] guy with the same rating?"

As with cultural training, most regional training is specific to predeployment efforts. It thus exhibits many of the same problems with measurement and tracking. Deployments, which are often the working proxy for regional and cultural experience for some interviewees, can be tracked by the services with personnel databases, but these databases are simply static monthly reports of which individuals are deployed during that month. Therefore, in order to determine the deployment history of any individual warfighter, an analyst would have to collect the monthly reports from a given area and pull out data for the individual service member month by month. As far as interviewees were aware, there is no centralized database for this information, and "the information would come in different pieces." An interviewee commented, "there's no RE [regional expertise] or [culture] test, at all, 'accepted' or not. There's no agreed-upon definition for RE." Given this lack of consistency in assessing regional skills, it is thus difficult to standardize any sort of tracking that could be used to assess capability or readiness in this area.

Although deployment history could potentially be used as a proxy for regional expertise or culture skills, there are many variations in missions, assignments, situations on the ground, and individual experiences. Thus, many interviewees felt that this history does not necessarily have any connection to an operationalized definition of *regional expertise*. Similarly, a college degree focused on a particular region may be outdated, broadly regional, or based on historical information, and a commander attempting to utilize regional expertise within his or her unit would most likely need current or more-localized information. As one interviewee commented, the college degree in regional studies or a similar subject area may be helpful to the individual in his or her career development but would probably not affect his or her performance in the field.

The greatest concern among interviewees about tracking any data was the quality of the data collected. In self-reports using the Interagency Language Roundtable scale, an individual may over- or underestimate his or her language skills without an actual DLPT. Moreover, without standardized guidelines for evaluating cultural and regional skills, there are likely to be serious discrepancies between the reporting from different services and COCOMs. Although most interviewees were optimistic about the LRI being developed for DRRS, that index will be useful only if the data in it are valid.

Provided that valid and standardized data could be obtained for the LRI or other tracking systems, there was some discrepancy among respondents as to what level of granularity would be most effective for planning purposes. Most interviewees expressed a preference for data with as much granularity as possible (e.g., the specific score a particular individual achieved on the DLPT on a certain date). However, some interviewees felt that, at their levels of responsibility (COCOM HQ), specific information was not as necessary. However, these interviewees did believe that, at operational and tactical levels, commanders would like to know information that is as specific as possible about their unit capabilities in order to properly plan and execute missions and training. Fields frequently requested included proficiency in English (because not all service members are native English speakers) or in specific dialects of a common language (e.g., Spanish).

Finally, a concern expressed by some interviewees was the disconnect between data (accurate or otherwise) and how those data are used to inform decisions. For example, one interviewee noted that one of his deputies was deployed as a political-military adviser and, upon arrival, was told that the position had been filled months prior. This interviewee commented, "We're not to the point yet where people are going to reach out and use the training and requirements we're trying to put on the table." Although not expressed as concisely by others, many interviewees identified concerns about operating tempo, shifting missions, and contingency operations as problems in assigning properly trained personnel or providing training to the units that require them.

CHAPTER FIVE

Conclusions and Recommendations

The objective of the research summarized in this report was to provide information to policy-makers about the available data to track LREC training and skills, as well as available information on how LREC affects readiness and mission accomplishment.

To do this, the research team addressed the following research questions:

1. According to the best available data, what is the relevance of LREC training and capabilities to overall unit readiness and mission accomplishment?
2. How does DoD currently track LREC training and capabilities of GPF?
3. To what extent does this tracking adequately reflect unit readiness and the ability to accomplish missions?
4. How can DoD improve tracking of LREC training and capabilities to adequately reflect unit readiness?

Research Question 1: According to the Best Available Data, What Is the Relevance of LREC Training and Capabilities to Overall Unit Readiness and Mission Accomplishment?

Practitioner and policymaker opinion, military lessons learned, and military directives emphasize the importance of language and culture proficiency. FM 3-24, authorship of which was overseen by GEN David Petraeus, states, "Successful conduct of COIN operations depends on thoroughly understanding the society and culture within which they are being conducted. [E]ffective COIN operations require a greater emphasis on certain skills, such as language and cultural understanding, than does conventional warfare" (DA, 2006, pp. 1-22, 1-23).

Business, industry, and the U.S. State Department have long provided expensive and time-consuming language and culture training for their employees going to foreign countries. A substantial number of anecdotes, as well as policy, suggest that LREC skills are essential to mission effectiveness. However, there have yet to be any rigorous, formal studies of whether LREC capabilities are linked to improved individual job performance or unit mission accomplishment in the military. Thus, more research is needed to confirm the relationship between LREC skills and mission accomplishment.

DoD has made several attempts to begin the process of collecting data through surveys, such as the SOF-A and a recent endeavor by the Marine Corps to interview redeploying marines. The research team's analysis of the SOF-A revealed that, although 40 percent of recently deployed service members positively rated the LREC training they received, slightly fewer said that it was beneficial to performing their job. Data limitations, as discussed in Chapter Three, prevent the team from inferring any causal link between the training these service members receive, job performance, and mission success.

A small but growing research literature uses data from service members with field experience to assess their perceptions of whether LREC training and skills are associated with job performance (e.g., Hardison et al., 2009; Zbylut et al., 2009). Although these studies suggest that some basic knowledge (e.g., common words and phrases, local norms and customs, appreciation of foreign cultures) is useful, they do not indicate that such knowledge is necessary for successfully performing a job. Further, this line of research does not establish a causal link between LREC and mission effectiveness.

Some of the interviewees in this study consider the impact of language and culture on mission readiness and on mission effectiveness to be too complex to measure except by survey or the general sense of the commanding officer. One interviewee observed that belief in the impact on mission effectiveness is a "love meter"—essentially, an estimate of how much commanders appreciate LREC skills. Given that predeployment training demands significantly exceed the time available for training, a possible indication of a commander's value of LREC may be the number of hours he or she allocates to LREC or the position of LREC training above and below other training priorities. However, given the diversity of operations and operational requirements, it would be difficult to compare data across missions.

The link from language and cultural expertise to mission readiness or effectiveness depends on multiple factors, which makes direct causal analysis difficult. The requirement for language and culture expertise may differ according to the type of operation and the specific locale. As one interviewee stated, "Readiness-wise, it doesn't play a big role unless you're talking about a specific mission or deployment." Interviewees stated that requirements for language and cultural expertise may also differ according to the COCOM or commander. Moreover, many factors other than GPF proficiency in language and culture affect mission effectiveness, from military strategy to global press.

Like *mission* and *mission effectiveness, language, regional expertise,* and *culture* are broad terms whose specific meanings may depend on specific contexts. For example, a language task may require a specific dialect, vocabulary, knowledge of certain phrases, and cultural information. That knowledge could have been taught in DLIFLC or university courses or learned in the home environment of a heritage speaker. Culture and regional expertise may be specific to a certain area or even to a certain social strata in a specific village, as well as to specific tasks. Cultural training may also be general (preparing a service member to go into any foreign environment) or specific (preparing the service member to go into a specific environment).

Considerable variation occurs in LREC requirements. Interviewees observed that requirements came from commanders and often changed with the commander. Without clear terms, clear requirements, and clear means of measuring LREC, identification of causal relationships cannot occur. Moreover, some interviewees pointed out that the issue is not necessarily the importance of LREC but its importance relative to other training, given the insufficient amount of total time available for training.

Research Question 2: How Does the U.S. Department of Defense Currently Track LREC Training and Capabilities of General Purpose Forces?

According to the interviewees, little systematic tracking of LREC training and capabilities of GPF has occurred. However, if LREC training is designed to bring the GPF to a very low but operationally effective level of training, there must be a mechanism in place to track that training, those skills, and the resulting operational effectiveness of the unit. None of those mechanisms is currently in place.

Interviewees reported that language is primarily tracked by DLPT scores and receipt of the FLPB. DLPT scores are general, multipurpose measures and do not provide information about a service member's ability to complete a specific task in a particular language or dialect. Further, not all service members with language expertise have taken the DLPT or qualify for an FLPB. Interviewees pointed out that they have not seen a clear correlation between DLPT scores (particularly at a very low level) and mission effectiveness or between DLPT scores and the ability to conduct tasks critical to the operations. Interviewees stated that cultural and regional training is often done by the unit, so commanders assume it to have been completed. Interviewees said that there is no tracking of these capabilities or training at the individual level.

As interviewees noted, each service collects some potentially relevant data (e.g., use of the language at home as a heritage speaker), but the services do not effectively track those data or make them available to unit commanders. As many interviewees pointed out, such data as home use of languages, university study, and prior deployments may indicate language and cultural or regional expertise. However, these interviewees also noted that such data do not always reliably predict such expertise. For instance, a person may have formerly been deployed to Iraq but have remained in the Green Zone and have had little opportunity to develop language and cultural or regional expertise. Even when these individual-level data are available, they are generally not available to unit commanders.

Requirement processes and requirements for LREC are still being defined by the Joint Chief's office. Using the requirements that are developed, one may be better able to determine the types of data needed to assess LREC capability. However, LREC expertise needed to achieve unit readiness and to accomplish missions is not yet well defined. Several interviewees pointed out that assessment of the impact of LREC capabilities on mission readiness is typically determined by the commanding officer and that the information used to make this determination is often at a "gut level" and not determined formally or systematically.

One particularly promising attempt to tie tracking of LREC training to mission readiness is currently under way in the Marine Corps. The Marine Corps has taken the approach of tracking LREC capabilities by *task* rather than by the DLPT (e.g., whether a marine can provide assistance in a medical situation). DLPT scores track general language skills, whereas task-based language skills are much more fine-grained. Tasks can then be more easily tied to mission readiness. However, assessment of readiness is primarily a "gut check" by the commander as to how his or her troops did in role-play exercises involving a wide range of military skills and is not empirically based.

Research Question 3: To What Extent Does This Tracking Adequately Reflect Unit Readiness and the Ability to Accomplish Missions?

It was the opinion of all interviewees that tracking of language and culture is insufficient to adequately reflect unit readiness and predict mission effectiveness. For DLO's purposes, it would be most useful to have a DoD-wide recommendation with respect to what constitutes mission readiness for GPF, similar to requirements for weapon training, equipment, and other readiness factors. However, although there are numerous concurrent studies about various aspects of LREC training and capabilities, making a "readiness recommendation" at this juncture would be premature and speculative at best. In order to answer the question of what LREC skills are necessary for GPF readiness, four critical and currently unanswered questions must be addressed:

- What is the causal relationship between LREC skills and capabilities and mission success?
- What data are needed to evaluate this potential causal relationship?
- How can that data be adequately tracked across the services?
- How can DoD adapt a portion of LREC training to the lowest common denominator of skills (either language, regional expertise, or cultural skills) necessary to accomplish missions, without overburdening both trainers and trainees?

The current efforts by JCS to establish requirements for each COCOM are a first step in developing testable causal relationships. Such requirements include the kinds of operations in which the COCOM is engaged or anticipates being engaged and the LREC skills required for each one. They also include a high-level description of the tasks within each operation and the languages and regions in which the operations will be conducted. DLIFLC's efforts to provide and evaluate task-based training upon unit request are a step in linking training to COCOM requirements, although such linkages are currently ad hoc.

Once DoD establishes linkages between task-based training and task-based effectiveness, the organization will need to do more work to distill this knowledge into the minimal training needed to meet the commander requirements for a particular mission. Although some capabilities and training may be fundamental to all missions and tasks, there will likely be other training that will be specific only to certain missions and tasks. More work is also needed in order to define the performance types and levels needed to achieve mission readiness, given the requirements of a particular mission.

In the following section, recommendations are provided for how to initiate a bottom-up—or task-based—LREC approach that will ultimately provide a top-down (e.g., DoD-wide) solution to the LREC readiness problem.

Research Question 4: How Can the U.S. Department of Defense Improve Tracking of LREC Training and Capabilities to Adequately Reflect Unit Readiness?

In this section, the research team presents recommendations based on the policy and directives and related academic literature that they reviewed, as well as the opinions provided to them in the interviews with LREC practitioners and policymakers. They first present recommendations that can be addressed in the short term. These recommendations are followed by longer-term suggestions for improving LREC tracking, as well as additional data-collection efforts that can be used to link LREC training and skills to mission readiness and mission effectiveness.

Recommendations

Short-Term Recommendations

Short-term recommendations include the following.

Recommendation 1: Standardize Terms Related to Language, Regional Expertise, and Culture

There is currently confusion about the term *LREC*, which is sometimes used to mean all of LREC and sometimes used to mean courses satisfying a military directive. As discussed, the services do not universally use the term *LREC*. There is similar confusion around the term *culture*, which is used sometimes to mean culture-general (i.e., training of how to be more open to learning about and interacting with other cultures with no specific culture designated) and other times culture-specific (i.e., training specific to one culture, such as that of a particular region) (see Abbe, Gulick, and Herman, 2007). Such terms as *linguist, translator*, and *interpreter* are often used variably. Developing a common vocabulary could be accomplished through discussions and agreements at management meetings and through providing the terminology via a DLO website that can be accessed by key stakeholders in the LREC world. These stakeholders include not only DoD and service-level representatives of the DLAP but also researchers who are conducting studies related to LREC. By standardizing LREC-related terms, DLO can begin to build a body of knowledge that builds on itself because the operationalization of key terms will be consistent.

Recommendation 2: Develop Measures of Mission Effectiveness

A key research question of this study was to determine the relevance of LREC training and capabilities to mission accomplishment and success. However, it is not yet possible to answer this question because it is not clear what it means to be mission effective or to achieve mission success. Assessing the linkage between LREC training and skills requires defining these terms, and the terms must be operationalized in a way on which key stakeholders agree. Workshops, such as those being conducted by JCS, could help to identify assessment approaches and measures. One approach may utilize focus groups of experts from the military and perhaps also from foreign partners to determine success metrics. This approach would provide a relatively quick way to get feedback regarding many complex factors.

The literature reviewed in this report has provided suggested recommendations for various measures, such as number of casualties or amount of anti-American graffiti. Such measures need to be used cautiously because they can be heavily influenced by factors other than language and culture expertise of GPF (e.g., by insurgent attacks). Tracking the local press to gauge local sentiment has also been recommended by returning commanders. There is now a rapidly emerging science and accompanying set of tools for sentiment analysis for situations in which local printed media are obtainable in machine-readable digital form in a language, such as Arabic, Dari, or Pashtu, that can be processed by machine translation or information-extraction tools. Such analysis enables the ingestion of large amounts of material and the identification of positive and negative sentiments about a particular topic (e.g., the United States or a U.S. policy).

There will likely not be a single measure that adequately reflects mission effectiveness. Moreover, the association between LREC training and skills and mission effectiveness and success will likely not be one-to-one. That is, many other factors will influence the course of missions and how successful they are or are not. In the next section are suggestions for how to collect these different pieces of information.

Recommendation 3: Develop Standardized After-Action Reports for Language, Regional Expertise, and Culture to Assess the Link Between LREC Training and Capabilities and Mission Success and Effectiveness

Standardized LREC AARs would collect data that could then be used to assess the association between LREC training and skills and mission success (as discussed in recommendation 2). DLO can collect this information across services, locations, and mission types. Such a systematic data-collection effort would provide quantitative data that could be analyzed to estimate the connection between LREC skills and capabilities and mission success. The standardized AAR should include fields pertaining to the mission (e.g., type, date, region or location), characteristics of the unit and its members (e.g., LREC training received, self-reported skills, heritage speakers), and metrics of mission effectiveness and success (e.g., commander ratings, objective measures). After these data have been collected for six months, they should be coded and analyzed both to provide an initial assessment of the LREC-mission success link and to determine whether and how the AAR should be edited or augmented to provide higher-quality data (e.g., what information was not collected that would be useful).

Ultimately, statistical analysis can be used to link these standardized AARs to objective measures of mission success determined in recommendation 2. For example, although it would be useful to ask unit commanders whether and how LREC skills of GPF aided in accomplishing their mission, it is also important to gauge success by objective measures. As noted earlier, some of those measures could include casualties (both military and civilian) or anti-American graffiti, for instance. These objective measures of success would balance any bias present in more-subjective measures, such as commander self-report.

Such AARs could provide data in the near term on how well GPF are meeting commander needs and expectations regarding LREC expertise. In terms of the Government Performance and Results Act (Pub. L. 103-62, 1993), the commander is the end user of the language and culture skills of GPF. According to the act, there should be continuing surveys to assess the end user's satisfaction with the services for which he or she stated a requirement. The surveys provide a feedback loop so that end users can adjust requirements as necessary. However, as noted earlier, commander assessments of satisfaction with training are subjective.

Recommendation 4: Develop Standardized Surveys for Deployed and Recently Deployed General Purpose Forces

As in recommendation 3, which is aimed at commanders, additional data that DoD could use to link LREC training and skills to mission readiness and success could come from recently deployed GPF service members. Such surveys could provide data on where skills need improvement. The services could collect data from GPF during and after deployment, as is being done now by DLIFLC. Note that the DLIFLC surveys focus on the effectiveness and applicability of the training DLIFLC provides.

Long-Term Recommendations
This section describes our long-term recommendations.

Recommendation 5: Develop an Infrastructure for LREC Data
This recommendation involves some subparts; however, the main goal of the recommendation is to develop a system in which LREC tracking, data collection, and analysis are standardized and widely available to interested parties. Specific recommendations are provided in this section.

First, take steps to *improve the accuracy of LREC data, especially those that may be self-reported.* In some sense, this issue will be addressed by recommendation 1; when LREC definitions are standardized, it should be clear what level of competency is necessary to fall into certain categories. However, because some interviewees noted that the association of DLPT scores and the ability of GPF to actually use LREC skills to perform tasks is unknown, we recommend further development of assessment studies that link DLPT scores to proficiency. It is understood that DLPT scores were not intended to convey information about ability to perform specific tasks. Yet, as noted by several interviewees, DLPT scores are currently being used to measure language proficiency. Either new assessments of skills are required (see recommendation 7) or a new study should assess the relationship between DLPT scores, especially at VLR, and the ability to complete needed tasks.

Second, because interview subjects alluded to problems with incomplete and uncertain data with respect to LREC training and capabilities among GPF, it may be helpful to *provide guidelines or information to aid interpretation to anyone using such data, including unit commanders and researchers.* Such documentation would include explicit information about the pros and cons of the data, as well as any caveats or exceptions that should be taken into account when using the data to make staffing decisions.

Finally, it would be helpful to *provide information about ongoing studies in order to facilitate communication across researchers, as well as LREC users and providers.* DLO planning documents, showing the relationships of the studies and development efforts, should be included in this effort. Given the fast-paced nature of the research in this field, it may be helpful to write consultant contracts such that the consultants can share drafts of the materials prior to final publication of those materials. Such a repository should, as much as possible, be made publicly available so that researchers across different disciplines can gain access to the latest LREC research available.

Recommendation 6: Develop a Causal Model Linking Language, Regional Expertise, and Culture to Mission Success
Ultimately, it will be necessary to establish a causal model linking LREC training and capabilities to mission effectiveness. Such an endeavor is designed to link specific types of training and skills to specific mission outcomes. Once this model is fully developed and tested, it will lead to recommendations about what it means to be mission ready in terms of LREC skills. Although there are multiple ways to develop a causal model, a bottom-up approach is recommended, based on smaller, related, linkable units of LREC skills and missions. This section offers an example of this bottom-up approach.

Both tracking and assessment become more manageable when they are based on specific requirements (e.g., the need for police action in a specific village where it is necessary to check identification, detaining people for further questioning). It is possible to develop a test of these individual skills. The services could then implement training if such skills were deficient. Thus, the impact of GPF LREC training and skills on unit readiness could be measured by whether GPF meet requirements set by the unit commander, as determined by testing of subcomponents of mission tasks. Eventually, subtasks can be aggregated until a baseline skill level is determined for the overarching mission, similar to the organization of the METL. However, some tasks (e.g., discussing a medical issue) may require specialized training in addition to baseline training.

One possible mechanism for linking tasks to LREC training and capabilities would be as follows:

- Define the operational missions, tasks, and subtasks (i.e., the requirements) and the relationships of those components. Much of this work is being done by JCS and PDRI.
- Document the language-, region-, and culture-specific requirements for each mission, task, and subtask. There is an effort by PDRI to provide this work breakdown; however, a copy of its work was not available for review at the time of this project.
- Develop comprehensive requirements for testing and training.
- Develop testing tied to the task requirements. Provide incentives or directives for service members to take the testing.
- Develop task-based training tied to the requirements of operational missions, tasks, and subtasks. Alternatively, map the extent to which the provided training covers the requirements.
- Provide feedback loops from unit commanders and officers about the extent to which the training is meeting LREC requirements for mission tasks and subtasks.
- Conduct or get inclusion in focus groups in-country with trusted partners to obtain feedback on adequate coverage and depth of LREC knowledge and to obtain insight about future requirements.
- Provide periodic analysis of requirements at a more specific level. For instance, are certain tasks or subtasks still required in a given language or dialect?
- Provide planning for requirements as far in advance as possible to allow for recruitment and training.
- Make deployment practices more predictable so that more training can be provided predeployment.
- Provide training to already-deployed troops, as is being done in many cases by DLIFLC.
- Improve practices to provide more consistency of requirements across changes of command.
- Identify skills that could be transferred or cross-trained for new requirements (e.g., for the same task in a different part of the country). Ensure that data are collected about these skills (e.g., a component of certain classes).

The end goal of such an endeavor is to (1) link LREC training and skills to mission accomplishment and (2) provide enough data to establish what it means to be LREC-ready. The approach outlined in this report is bottom-up in that it builds on individual tasks; however, once established, it will lend itself to the higher-level recommendations that DLO is seeking.

Developing a causal model between LREC training and skills and mission accomplishment or success will be a significant undertaking, requiring substantial resources (e.g., time, money, and manpower). A long-term investment will be needed. However, if DoD implements some of the suggested short-term recommendations, the process of model building can occur in small steps that build on one another.

Recommendation 7: Develop Tests of Training (i.e., Learning) That Are Associated with Skills That Have Been Linked to Mission Readiness

The previous recommendation suggested that DLO, with the help of data-collection efforts by the services, link training to operational missions (e.g. task-based training). Also, as noted in recommendation 4, surveys of GPF, either while in the field or after having just returned, may provide valuable insight on the effectiveness of LREC training. Survey developers might wish to consult the relevant recommendations in a recent report by the ARI on cross-cultural competence (Caligiuri et al., 2011). That report recommends that training and assessment be simultaneously developed to ensure that students are taught knowledge, skills, and attitudes that are accurately reflected in testing and assessment mechanisms (i.e., students are not tested on things that were not taught). As Caligiuri and colleagues note, "If the assessment design proceeds in isolation from cross-cultural training curriculum design, the Army may find itself with different messages about what it takes to be cross-culturally competent" (p. 52). If end minimum competency levels are not reflected in what is being tested, those tests are not very useful.

The same could be said of any type of training and assessment done by any of the services. The Caligiuri et al. report also recommends not mixing assessment efforts. Thus, "if the assessment has been designed and validity evidence gathered in relation to one purpose, that information is not necessarily informative about the utility of the assessment for another purpose" (p. 44). So that report reminds us that, if training assessments are designed to measure whether learning took place, those same assessments cannot necessarily tell us whether that same training will lead to actual changes in behavior or to organizational success.

In addition, the Caligiuri et al. report recommends designing appropriate assessment tools, taking into account the method of assessment (e.g., self-report, outside raters), the media used for assessment (e.g., paper and pencil, computer, interactive), the psychometric properties of the assessment tool (e.g., validity, reliability), the population being tested (e.g., individuals, units), and who administers and rates the assessment (e.g., administrative staff, subject-matter experts). It would also be useful to further systematize future data-collection efforts (e.g., through using Likert scales, objective measures of mission effectiveness) rather than just relying on anecdotal input so that results can be compared across groups and across time.

Summary of Recommendations

This section has provided recommendations aimed at improving tracking of LREC training and skills so that they better reflect readiness. Suggestions have been made for data-collection and analysis efforts to link LREC training and skills to mission effectiveness and success. Immediate recommendations include standardizing terms, such as *LREC* and *mission success*, as well as standardizing an LREC AAR. Long-term planning should include an effort to develop a strong infrastructure across LREC stakeholders such that information can easily be shared, to develop a theoretically sound causal model linking LREC skills to mission success, and to integrate task-based training and testing. The ultimate goal of these activities is to develop a set of readiness metrics, both at the general level for all GPF and at the mission-specific level, when specialized LREC skills may be required.

Policies and Directives Reviewed for This Analysis

For the analysis documented in this report, we reviewed the following policies:

- *Air Force Culture, Region, and Language Flight Plan*, May 2009 (U.S. Air Force, 2009)
- *Army Culture and Foreign Language Strategy*, December 1, 2009 (Department of the Army, 2009)
- *Center for Advanced Operational Cultural Learning (CAOCL) Key Leader Engagement Performance Assessment*, August 1, 2011 (CAOCL, 2011c)
- "COMISAF/USFOR-A Counterinsurgency (COIN) Training Guidance," November 10, 2009 (Headquarters, U.S. Forces–Afghanistan and International Security Assistance Force, 2009)
- Defense Language Program, DoDD 5160.41E, October 21, 2005 (DoD, 2010b)
- *Defense Language Transformation Roadmap*, January 2005 (DoD, 2005)
- *Strategic Plan for Language Skills, Regional Expertise, and Cultural Capabilities, 2011–2016* (DoD, 2011)
- *Language and Regional Expertise Planning*, CJCSI 3126.01, January 23, 2006 (CJCS, 2010)
- *U.S. Navy Language Skills, Regional Expertise, and Cultural Awareness Strategy*, January 2008 (Chief of Naval Operations, 2008)
- *Quadrennial Defense Review Report*, 2010 (DoD, 2010a)
- Secretary of Defense (SecDef) memo, "Implementing COIN Training Guidance to Support Execution of the President's Afghanistan-Pakistan Strategy," May 24, 2010 (Gates, 2010)
- *United States Marine Corps Center for Advanced Operational and Cultural Learning (CAOCL) Concept Document*, March 29, 2010 (CAOCL, 2010).

Interview List

Language, Regional Expertise, and Culture Users: Geographic Combatant Commands

- AFRICOM, May 10, 2011
- CENTCOM, July 26, 2011
- U.S. European Command (EUCOM, May 11, 2011
- USNORTHCOM, March 8, 2011
- U.S. Pacific Command (PACOM, March 21, 2011
- SOUTHCOM, April 14, 2011.

Language, Regional Expertise, and Culture Providers: Services

- U.S. Air Force, May 4, 2011
- U.S. Army, March 17, 2011
- Army National Guard (ARNG), May 3, 2011
- Marine Corps Intelligence Activity (MCIA), August 1, 2011
- Navy Foreign Language Office (NFLO), May 16, 2011.

Language, Regional Expertise, and Culture Providers: Other

- Defense Language Institute Foreign Language Center (DLIFLC), July 12, 2011
- Fort Polk, 162nd Infantry Brigade, April 18, 2011.

Language, Regional Expertise, and Culture Oversight and Tracking

- ARI, April 27, 2011
- DLO, March 25, 2011
- DMDC, March 7, 2011
- JCS, March 14, 2011
- Office of the Under Secretary of Defense for Intelligence (OUSD[I]), May 2, 2011.

Interview Questions

This appendix contains the text of the survey instrument.

Purposes of the Interviews

Our interviews had two purposes:

- Determine how LREC training and capabilities are tracked (e.g., individual, unit), captured, and pushed up for GPF and civilians.
- Determine the relevance of LREC training and capabilities among GPF and civilians to overall unit readiness and mission accomplishment.

List of Interview Questions for LREC Providers

1. Background
   a. What organization do you represent in answering these questions?
2. Personnel
   a. Among the deploying (or ground) forces, who receives LREC training?
   b. Among the deploying forces, who determines who gets LREC training?
      i. What criteria are used to determine who receives LREC training?
3. Curriculum
   a. Who determines LREC curriculum for deploying forces?
      i. What criteria are used to determine the content of training material?
   b. What metrics, if any, are used to determine [whether] the curriculum is effective?
   c. How static is the curriculum?
      i. If it does change, how often does it change?
      ii. Who determines [whether] a change is needed?
      iii. What criteria are used to determine [whether] a change is needed?
   d. To what extent is feedback from returning [noncommissioned officers] (NCOs) and [other] officers incorporated into LREC training for deploying forces?
   e. Are there any differences in the curriculum of LREC training provided to Reserve and Guard deploying forces? If so, what are they?

4. Follow-up
    a.   [With whom] else should we talk about LREC training and capabilities for deploying forces?

List of Interview Questions for Language, Regional Expertise, and Culture Users

1. Background
    a.   What organization do you represent in answering these questions?
2. Relevance of LREC capabilities for unit readiness and mission accomplishment
    a.   To what degree are language, regional, and cultural (LRC) capabilities relevant to overall unit readiness?
    b.   For what types of missions are LREC capabilities important?
        i.   What specific LREC skill sets are relevant for each type of mission?
        ii.   How specifically do those LREC skill sets aid in the accomplishment of those missions? Please give specific examples.
    c.   How do you assess the effectiveness of LREC training and capabilities among [GPF] for *mission readiness*?
    d.   How do you assess the effectiveness of LREC training and capabilities among [GPF] for *mission accomplishment*?
3. Capabilities and training
    a.   What criterion do you use to determine your command's [or] unit's LREC needs?
        i.   What type of skills have you requested?
            1.   To whom did you make this request?
            2.   How did you make this request?
        ii.   Did you get what you requested?
            1.   If not, what was the inadequacy? Can you give a specific example?
            2.   How did you handle your unfulfilled request or unmet need?
            3.   What impact did the inadequacy have on mission readiness or mission effectiveness?
    b.   In general, what LREC skills do the [GPF] have when they come to you?
        i.   What differences in skills have you witnessed based on deployment history?
        ii.   If predeployment LREC training was provided, did it match the mission the unit was asked to perform? Can you give an example?
            1.   If available, how did the unit use soldiers ([service members]) [who] received predeployment foreign-language training at [the] 0+ [or] 1 level and additional cultural training?
            2.   What changes to the predeployment training would you, as a commander, recommend?
    c.   What, if any, training does your command provide in LREC capabilities?
        i.   If training is available, is it once or recurrent?
    d.   What training is available to your troops from other sources (e.g., online, [temporary duty] [TDY]) for sustainment or implementation proficiency?

    e.   What recommendations would you make to improve the relevance, quality, and implementation of LREC training the [GPF] receive?

    f.   What opportunities exist for you or unit commanders to provide feedback about the training that ground forces receive?

4. Tracking

    a.   How do you track LREC *capabilities*, and at what level is that [tracking] done (e.g., individual, unit, component)?

        i.   Does this include self-reported competencies (e.g., individuals with preexisting LREC capabilities)?

    b.   How do you track LREC *training*, and at what level is that [tracking] done (e.g., individual, unit, component)?

    c.   How well does this current tracking of LREC training and capabilities adequately reflect unit readiness?

        i.   What should be tracked in terms of LREC training and capabilities to more accurately reflect unit readiness?

    d.   How do your organization and its components report on LREC training for [GPF]?

        i.   To whom do you report on LREC matters?

    e.   What problems do you see in the current tracking and reporting of LREC training or capabilities for [GPF]?

        i.   How would you improve on those problems?

    f.   What information would you want to have in order to know that a unit was language and culture mission-ready?

    g.   How would you and the commander like to see the information?

    h.   How reliable is current information?

5. Documentation

    a.   What documentation exists regarding the requirements, implementation, measurement, tracking, and impact of LREC training and capabilities for [GPF]?

        i.   Can these documents be made available to us?

    b.   What documentation exists regarding measurement of LREC training or capabilities as requirements for unit readiness?

        i.   How were those requirements generated?

6. Follow-up

    a.   [With whom] else should we talk about LREC capabilities, unit readiness, and mission accomplishment?

Five-Percent Confidence Intervals for the Status-of-Forces Analysis

CIs show the degree of uncertainty in the estimates. Because the SOF data use sampling statistics to select prospective respondents rather than a census framework in which all possible respondents are interviewed, there is some degree of error in estimating means, percentages, and other statistics. CIs represent the range of sample statistics (e.g., percentages) expected, given the sampling strategy used. The 95-percent CI reflects the range in which 95 percent of those sample statistics would fall.

Note that, for purposes of our tables, recent deployers are service members who have deployed at least once in the two years prior to the survey and have been deployed in OIF or OEF at least once since 9/11. All figures are percentages unless otherwise noted. CIs are presented in parentheses (lower bound, upper bound).

Table D.1
Deployment Since 9/11 to Operation Iraqi
Freedom or Operation Enduring Freedom

Personnel	Deployed	Not Deployed
Total Force	43.7 (42.3, 45.0)	56.3 (55.0, 57.7)
Officer	55.6 (53.9, 57.3)	44.4 (42.7, 46.1)
Enlisted	41.3 (39.7, 42.8)	58.7 (57.1, 60.3)

Table D.2
Type of Predeployment LREC Training Among Recent Deployers

Personnel	Language	Region	Culture	Any LRC
Total Force	27.6 (25.5, 29.8)	50.9 (48.7, 53.2)	59.9 (57.8, 62.1)	61.5 (59.4, 63.7)
Officer	23.3 (20.9, 25.7)	51.4 (48.8, 54.0)	62.5 (60.0, 64.9)	64.0 (61.6, 66.5)
Enlisted	29.0 (26.2, 31.7)	50.8 (47.9, 53.6)	59.2 (56.4, 61.9)	60.8 (58.1, 63.5)

Table D.3
Length of Predeployment LREC Training (in hours) Among Recent Deployers

Personnel	1–4	5–8	9–16	12–40	40+
Total Force	55.2 (52.2, 58.2)	24.5 (21.8, 27.2)	9.2 (7.2, 11.2)	5.9 (4.5, 7.3)	5.2 (3.6, 6.7)
Officer	53.4 (50.2, 56.6)	24.4 (21.6, 27.3)	8.7 (6.8, 10.6)	7.4 (5.6, 9.2)	6.0 (4.6, 7.5)
Enlisted	55.8 (52.0, 59.5)	24.6 (21.1, 28.0)	9.3 (6.8, 11.9)	5.4 (3.7, 7.2)	4.9 (2.9, 6.9)

Table D.4
Predeployment LREC Training, by Type of Trainer, Among Recent Deployers

Personnel	Service Culture Center	DLIFLC	Subject-Matter Expert	Trained Unit Trainer
Total Force	9.4 (7.8, 11.1)	5.8 (4.3, 7.4)	18.3 (16.0, 20.6)	66.4 (63.6, 69.3)
Officer	10.0 (8.2, 11.8)	5.4 (3.8, 7.0)	27.8 (24.8, 30.7)	56.8 (53.6, 60.0)
Enlisted	9.2 (7.2, 11.3)	6.0 (4.0, 7.9)	15.3 (12.5, 18.1)	69.5 (65.9, 73.0)

Table D.5
Location of Predeployment LREC Training Among Recent Deployers

Personnel	Classroom	Home Station	Distributed Learning or Online	Training Exercise	Other
Total Force	55.2 (52.1, 58.3)	24.5 (22.0, 27.0)	6.8 (5.7, 8.0)	5.9 (4.4, 7.4)	13.5 (11.1, 15.9)
Officer	55.5 (52.2, 58.7)	27.6 (24.7, 30.6)	6.7 (5.2, 8.3)	4.6 (3.4, 5.8)	10.1 (8.2, 12.1)
Enlisted	55.1 (51.1, 59.1)	23.5 (20.3, 26.6)	6.9 (4.3, 8.8)	6.2 (5.4, 8.3)	14.5 (11.5, 17.6)

Table D.6
Objectives of Predeployment LREC Training as Defined by Recent Deployers

Personnel	Geography	History	Cross-Cultural Competency Overview	Political Structure	Social Organizations	Religion	Economy	Negotiation
Total Force	87.7 (85.8, 89.6)	79.3 (76.9, 81.7)	81.1 (78.9, 83.3)	77.0 (74.4, 79.6)	83.2 (81.1, 85.4)	93.6 (92.0, 95.1)	72.9 (70.2, 75.6)	51.6 (48.5, 54.7)
Officer	88.8 (86.7, 90.8)	82.9 (80.0, 85.8)	78.1 (75.3, 80.9)	74.5 (71.4, 77.5)	79.3 (76.4, 82.1)	96.1 (94.9, 97.3)	69.3 (66.3, 72.4)	47.5 (44.3, 50.6)
Enlisted	87.4 (84.9, 89.8)	78.2 (75.1, 81.2)	82.0 (79.3, 84.7)	77.8 (74.5, 81.1)	84.4 (81.8, 87.1)	92.8 (90.8, 94.8)	74.0 (70.6, 77.4)	52.9 (49.0, 56.9)

Table D.7
Recent LREC Training Objectives Met and Inclusion of Supplemental Materials

Personnel	Training Met Objectives	Training Included Supplemental Material	Pocket Handbook or Reference Card	CD or DVD	Other
Total Force	88.5 (86.6, 90.3)	55.8 (52.9, 58.7)	86.0 (83.4, 88.6)	10.2 (8.0, 12.3)	3.8 (2.3, 5.4)
Officer	86.4 (84.1, 88.7)	54.5 (51.3, 57.7)	78.5 (75.0, 82.1)	16.4 (13.1, 19.6)	5.1 (3.5, 6.7)
Enlisted	89.1 (86.8, 91.4)	56.2 (52.6, 60.0)	88.3 (85.1, 91.4)	8.3 (5.6, 10.9)	3.4 (1.6, 5.3)

Table D.8
Satisfaction with Predeployment LREC Training Among Recent Deployers

Personnel	Very Satisfied	Satisfied	Neither	Dissatisfied	Very Dissatisfied
Total Force	5.4 (4.1, 6.6)	34.9 (32.0, 37.9)	46.9 (43.9, 49.9)	9.5 (7.5, 11.5)	3.3 (2.2, 4.4)
Officer	5.6 (4.1, 7.2)	36.0 (32.9, 39.1)	42.8 (39.6, 46.1)	12.0 (9.9, 14.1)	3.6 (2.3, 4.8)
Enlisted	5.3 (3.7, 6.8)	34.6 (30.9, 38.4)	48.2 (44.4, 52.0)	8.7 (6.2, 11.2)	3.2 (1.8, 4.6)

Table D.9
Job Performance Rating Associated with Predeployment LREC Training Among Recent Deployers

Personnel	Training Beneficial to Job Performance
Total Force	36.4 (33.2, 39.7)
Officer	33.0 (29.4, 36.5)
Enlisted	37.6 (33.4, 41.8)

References

Abbe, Allison, *Building Cultural Capability for Full-Spectrum Operations*, Arlington, Va.: U.S. Army Research Institute for the Behavioral and Social Sciences, Study Report 2008-04, January 2008. As of May 29, 2012:
http://www.au.af.mil/au/awc/awcgate/army/sr2008-04.pdf

Abbe, Allison, Lisa M. V. Gulick, and Jeffrey L. Herman, *Cross-Cultural Competence in Army Leaders: A Conceptual and Empirical Foundation*, Arlington, Va.: U.S. Army Research Institute for the Behavioral and Social Sciences, Study Report 2008-01, October 2007. As of May 29, 2012:
http://www.au.af.mil/au/awc/awcgate/army/sr2008-01.pdf

AFCLC—*See* Air Force Culture and Language Center.

Air Force Culture and Language Center, home page, updated May 23, 2012. As of May 30, 2012:
http://www.culture.af.mil/

Alliger, George M., and Elizabeth A. Janak, "Kirkpatrick's Levels of Training Criteria: Thirty Years Later," *Personnel Psychology*, Vol. 42, No. 2, June 1989, pp. 331–342.

Alliger, George M., Scott I. Tannenbaum, Winston Bennett Jr., Holly Traver, and Allison Shotland, "A Meta-Analysis of the Relations Among Training Criteria," *Personnel Psychology*, Vol. 50, No. 2, June 1997, pp. 341–358.

Alvarez, Kaye, Eduardo Salas, and Christina M. Garofano, "An Integrated Model of Training Evaluation and Effectiveness," *Human Resource Development Review*, Vol. 3, No. 2, December 2004, pp. 385–416.

Baker, COL Ralph O., U.S. Army, "The Decisive Weapon: A Brigade Combat Team Commander's Perspective on Information Operations," *Military Review*, Vol. 86, May–June 2010, pp. 13–32. As of May 29, 2012:
http://www.carlisle.army.mil/DIME/documents/Baker_Decisive%20Weapon.pdf

Becker, Gary S., *Human Capital: A Theoretical and Empirical Analysis, with Special Reference to Education*, New York: Columbia University Press, 1964.

Black, J. Stewart, "Work Role Transitions: A Study of American Expatriate Managers in Japan," *Journal of International Business Studies*, Vol. 19, No. 2, Summer 1988, pp. 277–294.

Black, J. Stewart, and Hal B. Gregersen, "Antecedents to Cross-Cultural Adjustment for Expatriates in Pacific Rim Assignments," *Human Relations*, Vol. 44, No. 5, May 1991, pp. 497–515.

Black, J. Stewart, and Mark Mendenhall, "Cross-Cultural Training Effectiveness: A Review and a Theoretical Framework for Future Research," *Academy of Management Review*, Vol. 15, No. 1, January 1990, pp. 113–136.

Brewster, Chris, "Effective Expatriate Training," in Jan Selmer, ed., *Expatriate Management: New Ideas for International Business*, Westport, Conn.: Quorum Books, 1995, pp. 57–71.

Brislin, Richard W., and Tomoko Yoshida, *Intercultural Communication Training: An Introduction*, Thousand Oaks, Calif.: Sage Publications, 1994.

Caligiuri, Paula, R. Noe, R. Nolan, A. M. Ryan, and F. Drasgow, "Training, Developing, and Assessing Cross-Cultural Competence in Military Personnel," Arlington, Va.: U.S. Army Research Institute for the Behavioral and Social Sciences, Technical Report 1284, 2011.

CAOCL—*See* Center for Advanced Operational Culture Learning.

Center for Advanced Operational Culture Learning, *United States Marine Corps Center for Advanced Operational and Cultural Learning (CAOCL) Concept Document*, March 29, 2010.

———, *CAOCL Culture and Language Survey: Importance of Culture Versus Language*, Quantico, Va.: U.S. Marine Corps, 2011a.

———, *CAOCL Language and Culture Survey: Ground Combat Arms Responses*, Quantico, Va.: U.S. Marine Corps, 2011b.

———, *Center for Advanced Operational Cultural Learning (CAOCL) Key Leader Engagement Performance Assessment*, August 1, 2011c.

Chairman of the Joint Chiefs of Staff, *Language and Regional Expertise Planning*, Washington, D.C., Instruction 3126.01, January 23, 2006, incorporating change 1, April 14, 2006, directive current as of November 27, 2010. As of May 29, 2012:
http://www.dtic.mil/cjcs_directives/cdata/unlimit/3126_01.pdf

Chief of Naval Operations, *U.S. Navy Language Skills, Regional Expertise and Cultural Awareness Strategy*, Washington, D.C., January 2008. As of May 29, 2012:
http://handle.dtic.mil/100.2/ADA503388

Cognitive Performance Group, *Culture Knowledge and Survival Language Skill Pre-Deployment Training Project: Stage 1*, Patrick Air Force Base, Fla.: Defense Equal Opportunity Management Institute, technical report, 2010.

———, *Culture Knowledge and Survival Language Skill Pre-Deployment Training Project: Phase II Final Report*, Arlington, Va.: Defense Language Office, March 15, 2011. As of May 29, 2012:
http://issuu.com/cognition/docs/phase_ii_final_research_report_15_march_2011

DA—*See* Department of the Army.

Defense Language Institute Foreign Language Center, Defense Language Proficiency Test Program, *Guide to Educational Credit by Examination*, August 2007. As of October 5, 2011:
http://www.dliflc.edu/archive/documents/DLPT_Credit_by_Exam_Policy.pdf

Defense Manpower Data Center, *June 2010 Status of Forces Survey of Active-Duty Members: Administration, Datasets, and Codebook*, Arlington, Va., Report 2009-074, 2010a.

———, *June 2010 Status of Forces Survey of Active-Duty Members: Statistical Methodology Report*, Arlington, Va., Report 2009-075, 2010b.

Department of the Army, *Counterinsurgency*, Washington, D.C., Field Manual 3-24, December 2006. As of May 29, 2012:
http://www.fas.org/irp/doddir/army/fm3-24.pdf

———, *Army Culture and Foreign Language Strategy*, December 1, 2009.

Department of the Navy, *Operational Culture and Language Training and Readiness Manual*, Washington, D.C.: Headquarters, U.S. Marine Corps, NAVMC 3500.65, April 8, 2009.

Deshpande, Satish P., and Chockalingam Viswesvaran, "Is Cross-Cultural Training of Expatriate Managers Effective? A Meta-Analysis," *International Journal of Intercultural Relations*, Vol. 16, No. 3, Summer 1992, pp. 295–310.

DMDC—*See* Defense Manpower Data Center.

DoD—*See* U.S. Department of Defense.

EduMetrica, *Framework for the Very Low Range Defense Language Proficiency Tests (VLR DLPT)*, version 5, Washington, D.C., March 2010.

Flynn, MG Michael T., U.S. Army, Capt Matt Pottinger, U.S. Marine Corps, and Paul D. Batchelor, *Fixing Intel: A Blueprint for Making Intelligence Relevant in Afghanistan*, Washington, D.C.: Center for a New American Security, January 2010. As of May 29, 2012:
http://www.cnas.org/files/documents/publications/AfghanIntel_Flynn_Jan2010_code507_voices.pdf

GAO—*See* U.S. Government Accountability Office.

Gates, Robert, Secretary of Defense, "Implementing Counterinsurgency (COIN) Training Guidance to Support Execution of the President's Afghanistan-Pakistan Strategy," memorandum for secretaries of the military departments, chair of the Joint Chiefs of Staff, under secretaries of defense, commanders of the combatant commands, commander of U.S. Forces–Afghanistan, general counsel of the U.S. Department of Defense, the director of cost assessment and program evaluation, and the directors of the defense agencies, May 24, 2010.

Gregersen, Hal B., and J. Stewart Black, "Antecedents to Commitment to a Parent Company and a Foreign Operation," *Academy of Management Journal*, Vol. 35, No. 1, March 1992, pp. 65–90.

Gudykunst, William B., Ruth M. Guzley, and Mitchell R. Hammer, "Designing Intercultural Training," in Dan Landis and Rabi S. Bhagat, eds., *Handbook of Intercultural Training*, 2nd ed., Thousand Oaks, Calif.: Sage Publications, 1995, pp. 61–80.

Hardison, Chaitra M., Carra S. Sims, Farhana Ali, Andres Villamizar, Benjamin F. Mundell, and Paul Howe, *Cross-Cultural Skills for Deployed Air Force Personnel: Defining Cross-Cultural Performance*, Santa Monica, Calif.: RAND Corporation, MG-811-AF, 2009. As of May 30, 2012:
http://www.rand.org/pubs/monographs/MG811.html

Harris, Philip R., and Robert T. Moran, *Managing Cultural Differences*, Houston: Gulf Publishing Co., 1996.

Headquarters, U.S. Marine Corps, *Marine Corps Language, Regional, and Culture Strategy: 2010–2015*, Washington, D.C., Navy and Marine Corps Directive HQ 335, 2011.

Headquarters, USMC—*See* Headquarters, U.S. Marine Corps.

Headquarters, U.S. Forces–Afghanistan and International Security Assistance Force, "COMISAF/USFOR-A Counterinsurgency (COIN) Training Guidance," memorandum, Kabul, Afghanistan, November 10, 2009. As of May 30, 2012:
http://usacac.army.mil/cac2/coin/repository/COMISAF_COIN_Training_Guidance.pdf

Hemmasi, Masoud, Meredith Downes, and Iris I. Vamer, "An Empirically-Derived Multidimensional Measure of Expatriate Success: Reconciling the Discord," *International Journal of Human Resource Management*, Vol. 21, No. 7, June 2010, pp. 982–998.

Kealey, Daniel J., and David R. Protheroe, "The Effectiveness of Cross-Cultural Training for Expatriates: An Assessment of the Literature on the Issue," *International Journal of Intercultural Relations*, Vol. 20, No. 2, Spring 1996, pp. 141–165.

Keen, LTG P. K. (Ken), LTC Matthew G. Elledge, LTC Charles W. Nolan, and LTC Jennifer L. Kimmey, U.S. Army, "Foreign Disaster Response: Joint Task Force–Haiti Observations," *Military Review*, November–December 2010, pp. 85–96. As of May 30, 2012:
http://usacac.army.mil/CAC2/MilitaryReview/Archives/English/MilitaryReview_20101231_art015.pdf

Kuehl, LTC Dale, U.S. Army, "Testing Galula in Ameriyah: The People Are the Key," *Military Review*, March–April 2009, pp. 72–80. As of May 30, 2012:
http://usacac.army.mil/CAC2/MilitaryReview/Archives/English/MilitaryReview_20090430_art012.pdf

Kirkpatrick, Donald, "Techniques for Evaluating Training Programs," *Journal of ASTD*, Vol. 13, 1959a, pp. 3–9.

———, "Techniques for Evaluating Training Programs: Part 2—Learning," *Journal of ASTD*, Vol. 13, 1959b, pp. 21–26.

———, "Techniques for Evaluating Training Programs: Part 3—Behavior," *Journal of ASTD*, Vol. 14, 1960a, pp. 13–18.

———, "Techniques for Evaluating Training Programs: Part 4—Results," *Journal of ASTD*, Vol. 14, 1960b, pp. 28–32.

Lewis, M. Paul, ed., *Ethnologue: Languages of the World*, 16th ed., Dallas: SIL International, 2009. As of October 5, 2011:
http://www.ethnologue.com/

Mendenhall, Mark E., Ina Ehnert, Torsten M. Kühlmann, Gary Oddou, Joyce S. Osland, and Günter K. Stahl, "Evaluation Studies of Cross-Cultural Training Programs: A Review of the Literature from 1988–2000," in Dan Landis, Janet Marie Bennett, and Milton J. Bennett, eds., *Handbook of Intercultural Training*, Thousand Oaks, Calif.: Sage Publications, 2004, pp. 129–144.

Murphy, Dennis M., "In Search of the Art and Science of Strategic Communication," *Parameters*, Winter 2009–2010, pp. 105–116. As of May 30, 2012:
http://www.au.af.mil/au/awc/awcgate/parameters/murphy.pdf

Norton-Taylor, Richard, "London Theatre Troupe to Perform Play on Afghan History for U.S. Military," *Guardian*, January 9, 2011. As of September 22, 2011:
http://www.guardian.co.uk/world/2011/jan/09/london-troupe-pentagon-afghanistan

Public Law 103-62, Government Performance and Results Act, August 3, 1993.

Puck, Jonas, Markus Kittler, and Christopher Wright, "Does It Really Work? Re-Assessing the Impact of Pre-Departure Cross-Country Training on Expatriate Adjustment," *International Journal of Human Resource Management*, Vol. 19, No. 12, 2008, pp. 2182–2197.

U.S. Air Force, *Air Force Culture, Region and Language Flight Plan*, May 2009. As of May 30, 2012:
http://www.culture.af.mil/library/pdf/cultureflightplan.pdf

U.S. Department of Defense, Policy on Graduate Education for Military Officers, Directive 1322.10, August 31, 1990.

———, *Statistical Design of the Status of Forces Surveys of Active-Duty Members*, Arlington, Va., Report 2002-033, 2002.

———, *Defense Language Transformation Roadmap*, Washington, D.C., January 2005. As of May 29, 2012:
http://www.defense.gov/news/Mar2005/d20050330roadmap.pdf

———, Foreign Language Proficiency Bonus (FLPB), Washington, D.C., Instruction 7280.03, August 20, 2007. As of May 29, 2012:
http://www.culture.af.mil/library/pdf/728003.pdf

———, *Quadrennial Defense Review Report*, Washington, D.C., February 2010a. As of May 29, 2012:
http://www.defense.gov/qdr/

———, Defense Language Program (DLP), Washington, D.C., Directive 5160.41E, October 21, 2005, incorporating change 1, May 27, 2010b. As of May 29, 2012:
http://www.dtic.mil/whs/directives/corres/pdf/516041p.pdf

———, *Strategic Plan for Language Skills, Regional Expertise, and Cultural Capabilities, 2011–2016*, Washington, D.C., 2011.

U.S. Government Accountability Office, *Military Training: Actions Needed to Improve Planning and Coordination of Army and Marine Corps Language and Culture Training—Report to Congressional Committees*, Washington, D.C., GAO-11-456, May 2011a. As of May 29, 2012:
http://www.gao.gov/new.items/d11456.pdf

———, *Language and Culture Training: Opportunities Exist to Improve Visibility and Sustainment of Knowledge and Skills in Army and Marine Corps General Purpose Forces—Report to Congressional Committees*, Washington, D.C., GAO-12-50, October 2011b. As of May 29, 2012:
http://purl.fdlp.gov/GPO/gpo16097

U.S. House of Representatives Committee on Armed Services Subcommittee on Oversight and Investigations, *Building Language Skills and Cultural Competencies in the Military: DoD's Challenge in Today's Educational Environment*, Washington, D.C., November 2008. As of May 29, 2012:
http://cgsc.cdmhost.com/cdm/singleitem/collection/p4013coll11/id/1394/rec/2

———, *Building Language Skills and Cultural Competencies in the Military: Bridging the Gap*, Washington, D.C., December 2010. As of May 29, 2012:
http://armedservices.house.gov/index.cfm/files/serve?File_id=2361fa65-7e40-41df-8682-9725d1c377da

Vogel, A. J., J. J. Van Vuuren, and S. M. Millard, "Preparation, Support and Training Requirements of South African Expatriates," *South African Journal of Business Management*, Vol. 39, No. 3, 2008, pp. 33–40.

Zbylut, Michelle Ramsden, LTC Brandon McGowan, Joint Center for International Security Force Assistance, MSG Michael Beemer, Joint Center for International Security Force Assistance, Jason M. Brunner, and Christopher L. Vowels, *The Human Dimension of Advising: An Analysis of Interpersonal, Linguistic, Cultural, and Advisory Aspects of the Advisor Role*, Fort Leavenworth, Kan.: U.S. Army Research Institute for the Behavioral and Social Sciences, Technical Report 1248, June 2009. As of May 30, 2012:
http://www.dtic.mil/cgi-bin/GetTRDoc?AD=ADA507713